Everyday Mathematics®

The University of Chicago School Mathematics Project

Student Math Journal
Volume 2

Grade **1**

DAKOTA

**Wright Group**

The McGraw·Hill Companies

The University of Chicago School Mathematics Project (UCSMP)

Max Bell, Director, UCSMP Elementary Materials Component; Director, *Everyday Mathematics* First Edition
James McBride, Director, *Everyday Mathematics* Second Edition
Andy Isaacs, Director, *Everyday Mathematics* Third Edition
Amy Dillard, Associate Director, *Everyday Mathematics* Third Edition

Authors

Max Bell	Robert Hartfield	Kathleen Pitvorec
Jean Bell	Andy Isaacs	Peter Saecker
John Bretzlauf	James McBride	
Amy Dillard	Rachel Malpass McCall*	

*Third Edition only

Technical Art
Diana Barrie

Teachers in Residence
Jeanine O'Nan Brownell
Andrea Cocke
Brooke A. North

Editorial Assistant
Rossita Fernando

Contributors

Cynthia Annorh, Robert Balfanz, Judith Busse, Mary Ellen Dairyko, Lynn Evans, James Flanders, Dorothy Freedman, Nancy Guile Goodsell, Pam Guastafeste, Nancy Hanvey, Murray Hozinsky, Deborah Arron Leslie, Sue Lindsley, Mariana Mardrus, Carol Montag, Elizabeth Moore, Kate Morrison, William D. Pattison, Joan Pederson, Brenda Penix, June Ploen, Herb Price, Dannette Riehle, Ellen Ryan, Marie Schilling, Susan Sherrill, Patricia Smith, Kimberli Sorg, Robert Strang, Jaronda Strong, Kevin Sweeney, Sally Vongsathorn, Esther Weiss, Francine Williams, Michael Wilson, Izaak Wirzup

Photo Credits
©Ralph A. Clevenger/CORBIS, cover, *center;* Getty Images, cover, *bottom left;* ©Tom and Dee Ann McCarthy/CORBIS, cover *right.*

www.WrightGroup.com

Printed in the United States of America.

Send all inquiries to:
Wright Group/McGraw-Hill
P.O. Box 812960
Chicago, IL 60681

ISBN 0-07-604536-6

16 CPC 12 11 10

The *McGraw·Hill* Companies

Contents

UNIT 6 Developing Fact Power

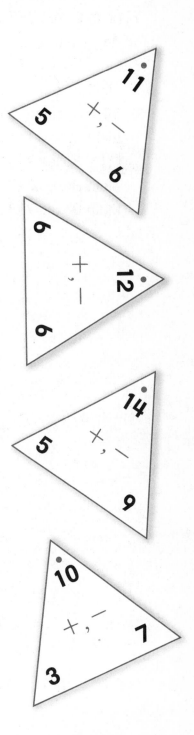

UNIT 7 Geometry and Attributes

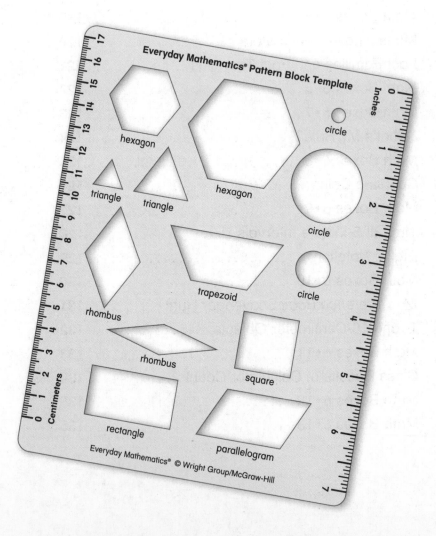

Mental Arithmetic, Money, and Fractions

UNIT 9 Place Value and Fractions

UNIT 10 End-of-Year Reviews and Assessments

Activity Sheets

LESSON 6·1 Dice-Throw Record 2

Unit
dice
dots

Record each fact and its turn-around fact once.

											12
											11
											10
									6 + 3		9
											8
											7
											6
						1 + 4					5
						4 + 1					
											4
											3
											2

LESSON 6·1   **Math Boxes**

1. Write the sums.

$2 + 4 =$ _6_

_5_ $= 3 + 2$

_6_ $= 1 + 5$

$4 + 0 =$ _4_

2. What is the missing rule?

in → Rule → out

in	out
3	5
17	19
14	16

Fill in the circle next to the best answer.

Ⓐ +3   Ⓑ +2

Ⓒ −2   Ⓓ −5

3. Add.

$3 + 4 =$ _7_

$4 + 3 =$ _7_

$2 + 7 =$ _9_

$7 + 2 =$ _9_

4. Draw lines to match the shapes that look alike.

Column A   Column B

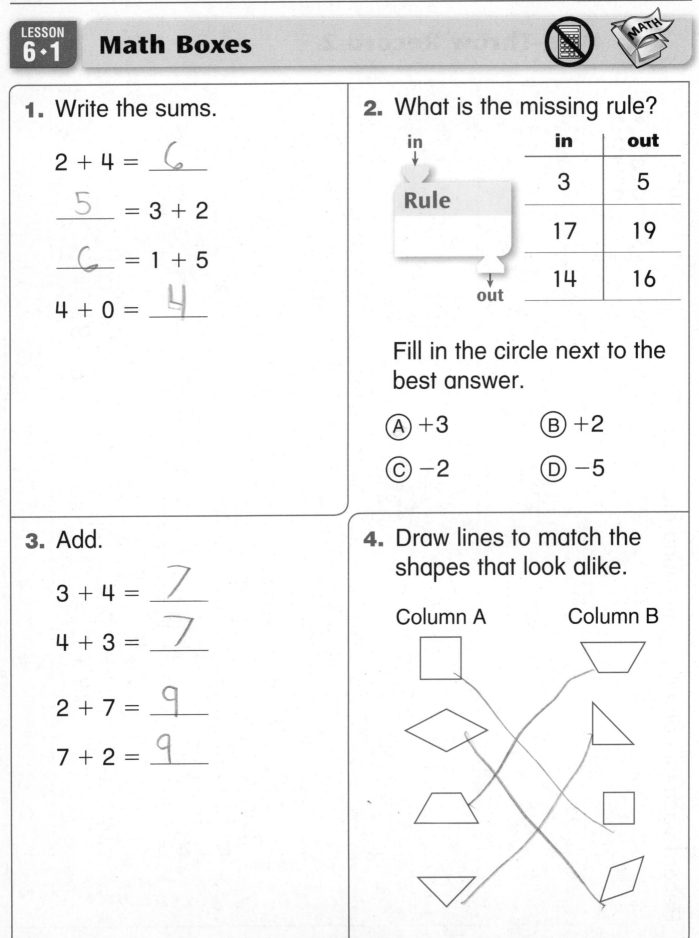

Date _____

1. Write other names for 11.

11

8 + 3

10 + 1

13 − 2

2. Write other names for 12.

12

1 dozen

~HHt~ ~HHt~ //

3 + 3 + 3 + 3

15 − 3

3. Cross out the names that don't belong in the 10-box.

10

~HHt~ ~HHt~

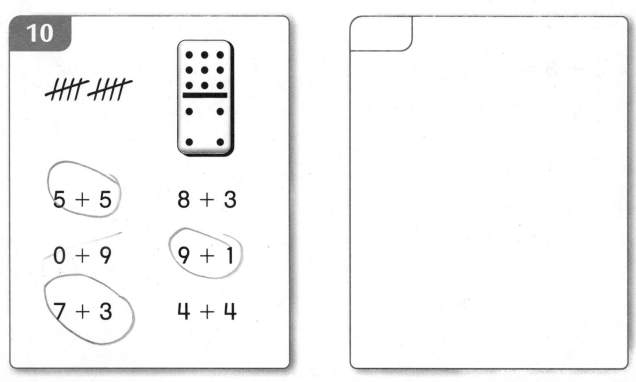

⟨5 + 5⟩   8 + 3

0 + 9   ⟨9 + 1⟩

⟨7 + 3⟩   4 + 4

4. Make your own.

Date _____

Math Boxes

1. Write 5 more names for 10.

10

5 + 5

ten

HHT HHT

2. Solve the riddles.

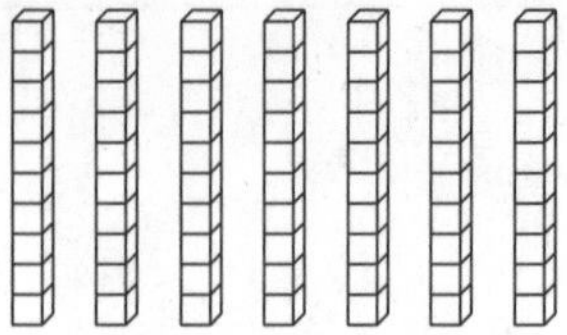

What am I? _____

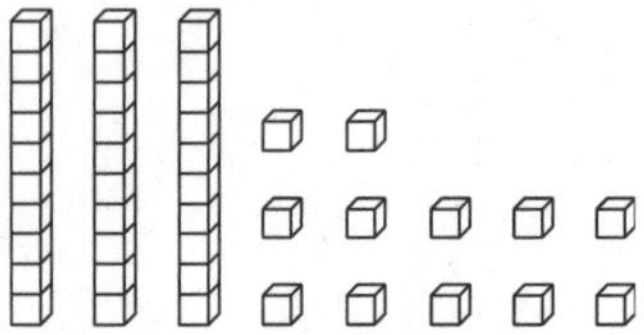

What am I? _____

3. Add.

2 + 2 = 4

3 + 3 = 6

4 + 4 = 8

5 + 5 = 10

4. Shade the biggest triangle.

LESSON 6·3 Fact Families

Write the 3 numbers for each domino.
Use the numbers to write the fact family.

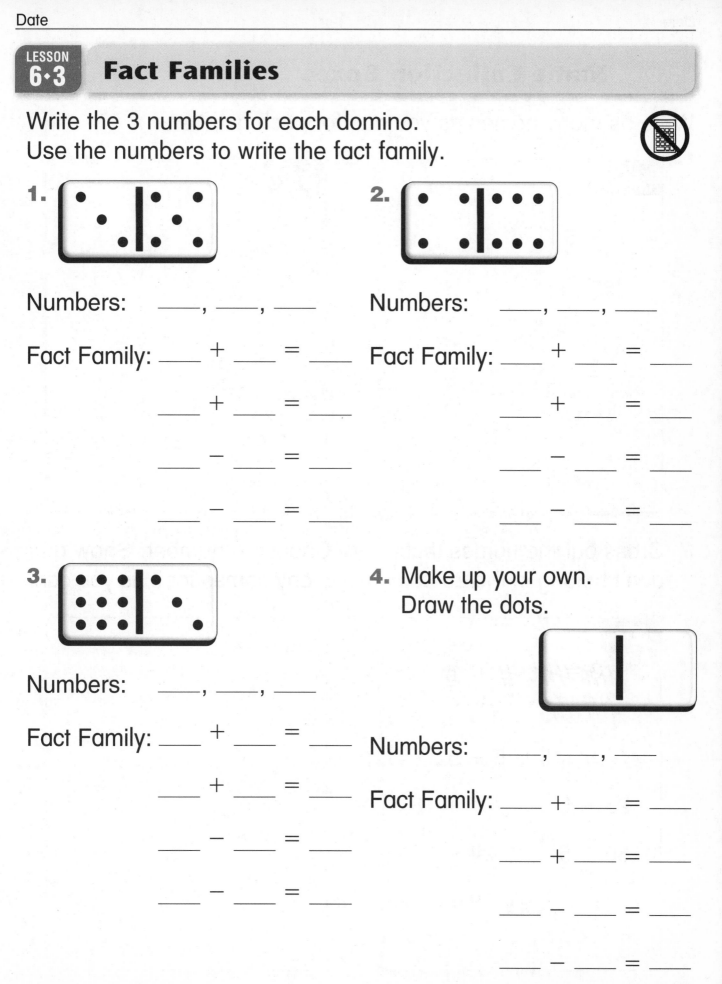

1.

Numbers: ____, ____, ____

Fact Family: ____ + ____ = ____

_____ + ____ = ____

_____ − ____ = ____

_____ − ____ = ____

2.

Numbers: ____, ____, ____

Fact Family: ____ + ____ = ____

_____ + ____ = ____

_____ − ____ = ____

_____ − ____ = ____

3.

Numbers: ____, ____, ____

Fact Family: ____ + ____ = ____

_____ + ____ = ____

_____ − ____ = ____

_____ − ____ = ____

4. Make up your own.
Draw the dots.

Numbers: ____, ____, ____

Fact Family: ____ + ____ = ____

_____ + ____ = ____

_____ − ____ = ____

_____ − ____ = ____

LESSON 6·3 Name-Collection Boxes

Write as many names as you can for each number.

1.

13

2.

20

3. Cross out the names that don't belong in the 25 box.

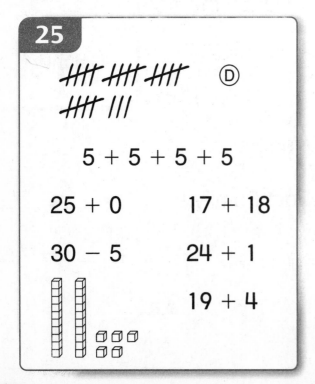

4. Choose a number. Show as many names for it as you can.

Date

1. Write the sums.

$1 + 8 =$ _9_

_8_ $= 6 + 2$

$$
\begin{array}{cc}
4 & 7 \\
+\,2 & +\,0 \\
\hline
6 & 7 \\
\end{array}
$$

2. Fill in the missing numbers.

in ↓

Rule

$+10$

out ↓

in	out
13	
18	
79	
93	
125	

3. Add.

$5 + 4 =$ _9_

$4 + 5 =$ _9_

$2 + 3 =$ _5_

$3 + 2 =$ _5_

4. Draw lines to match the shapes that look alike.

Column A   Column B

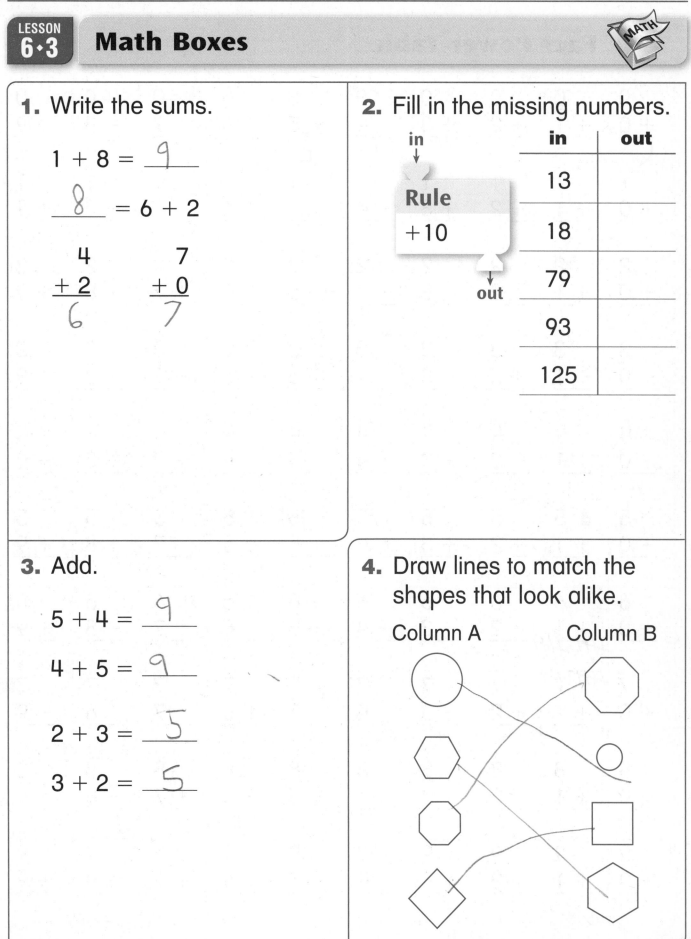

LESSON 6·4 Fact Power Table

0 +0 *0*	0 +1 *1*	0 +2 *2*	0 +3 *3*	0 +4 *4*	0 +5 *5*	0 +6 *6*	0 +7 *7*	0 +8 *8*	0 +9 *9*
1 +0 *1*	1 +1 *2*	1 +2 *3*	1 +3 *4*	1 +4 *5*	1 +5 *6*	1 +6 *7*	1 +7 *8*	1 +8 *9*	1 +9 *10*
2 +0 *2*	2 +1 *3*	2 +2 *4*	2 +3 *5*	2 +4	2 +5	2 +6	2 +7	2 +8	2 +9
3 +0	3 +1	3 +2	3 +3	3 +4	3 +5	3 +6	3 +7	3 +8	3 +9
4 +0	4 +1	4 +2	4 +3	4 +4	4 +5	4 +6	4 +7	4 +8	4 +9
5 +0	5 +1	5 +2	5 +3	5 +4	5 +5	5 +6	5 +7	5 +8	5 +9
6 +0	6 +1	6 +2	6 +3	6 +4	6 +5	6 +6	6 +7	6 +8	6 +9
7 +0	7 +1	7 +2	7 +3	7 +4	7 +5	7 +6	7 +7	7 +8	7 +9
8 +0	8 +1	8 +2	8 +3	8 +4	8 +5	8 +6	8 +7	8 +8	8 +9
9 +0	9 +1	9 +2	9 +3	9 +4	9 +5	9 +6	9 +7	9 +8	9 +9

Date

Math Boxes

1. Label the box.
Add 5 names.

HHT HHT HHT   20 − 5

7 + 8

2. What is the number?

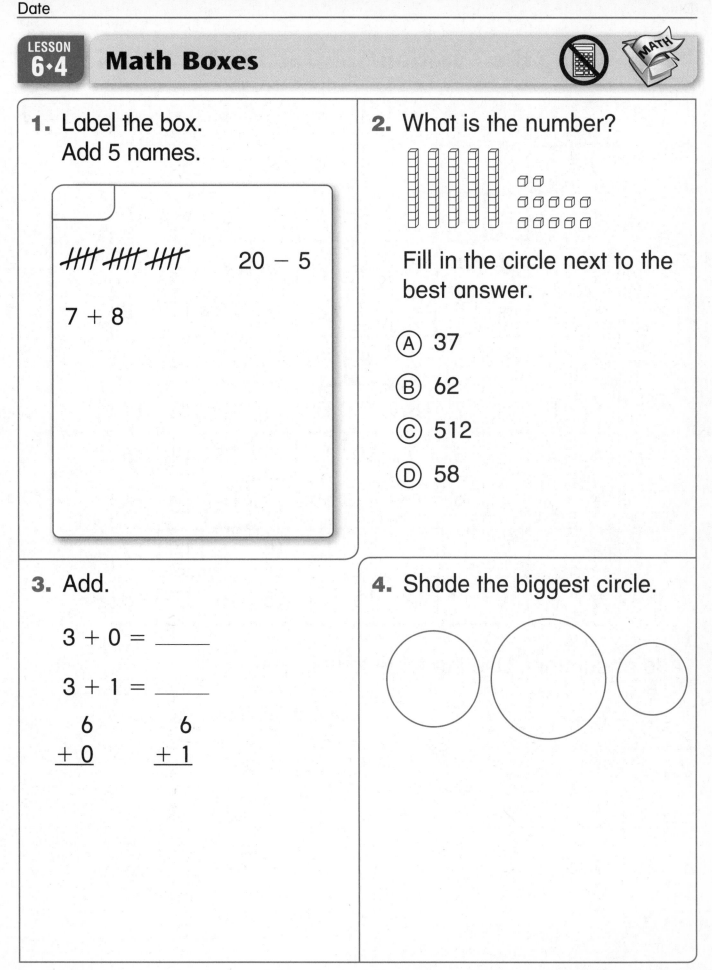

Fill in the circle next to the
best answer.

Ⓐ 37

Ⓑ 62

Ⓒ 512

Ⓓ 58

3. Add.

3 + 0 = _____

3 + 1 = _____

```
  6        6
+ 0      + 1
```

4. Shade the biggest circle.

LESSON 6·5 **Using the Addition/Subtraction Facts Table 2**

+,−	0	1	2	3	4	5	6	7	8	9
0	0	1	2	3	4	5	6	7	8	9
1	1	2	3	4	5	6	7	8	9	10
2	2	3	4	5	6	7	8	9	10	11
3	3	4	5	6	7	8	9	10	11	12
4	4	5	6	7	8	9	10	11	12	13
5	5	6	7	8	9	10	11	12	13	14
6	6	7	8	9	10	11	12	13	14	15
7	7	8	9	10	11	12	13	14	15	16
8	8	9	10	11	12	13	14	15	16	17
9	9	10	11	12	13	14	15	16	17	18

Add or subtract. Use the table to help you.

1. $5 + 6 =$ _____

2. $11 - 5 =$ _____

3. $8 + 4 =$ _____

4. $12 - 4 =$ _____

5. $7 + 8 =$ _____

6. $15 - 8 =$ _____

7. $9 + 9 =$ _____

8. $18 - 9 =$ _____

9. $9 + 7 =$ _____

10. $16 - 9 =$ _____

LESSON 6·5 — Math Boxes

1. Subtract.

$5 - 1 =$ _____

$4 - 2 =$ _____

_____ $= 6 - 0$

_____ $= 3 - 3$

2. Write the fact family.

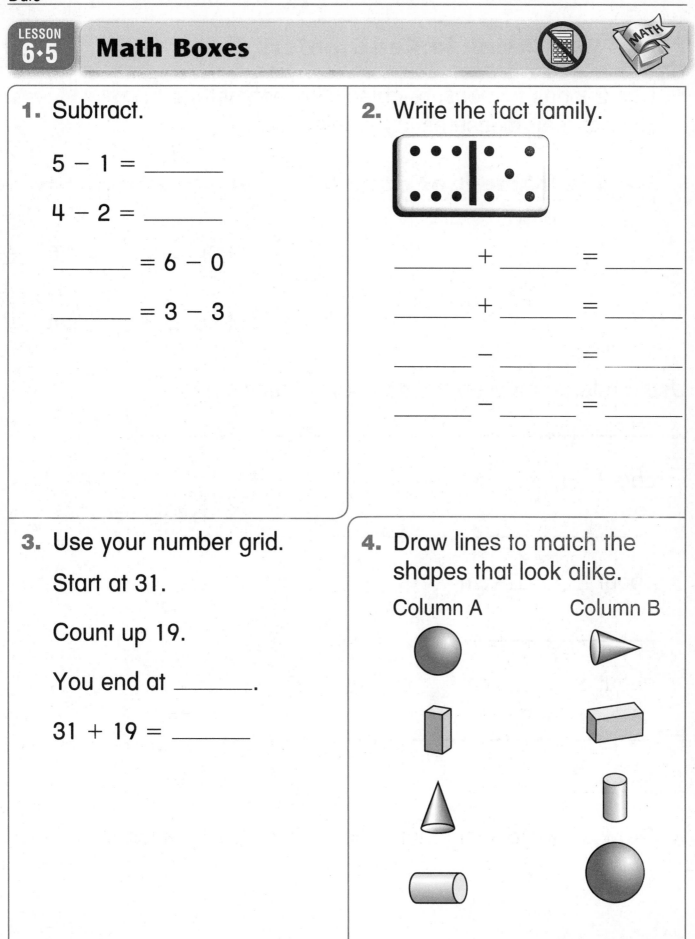

_____ $+$ _____ $=$ _____

_____ $+$ _____ $=$ _____

_____ $-$ _____ $=$ _____

_____ $-$ _____ $=$ _____

3. Use your number grid.

Start at 31.

Count up 19.

You end at _____.

$31 + 19 =$ _____

4. Draw lines to match the shapes that look alike.

Column A   Column B

LESSON 6·6 Measuring in Centimeters

1. Use 2 longs to measure objects in centimeters.
Record their measures in the table.

Object (name it or draw it)	My measurement
	about _____ cm
	about _____ cm

Use a ruler to measure to the nearest centimeter.

2. _____

about _____ cm

3. _____

about _____ cm

4. _____

about _____ cm

5. _____

about _____ cm

6. Draw a line segment that is about 9 centimeters long.

Date _____

Write the fact family for each Fact Triangle.

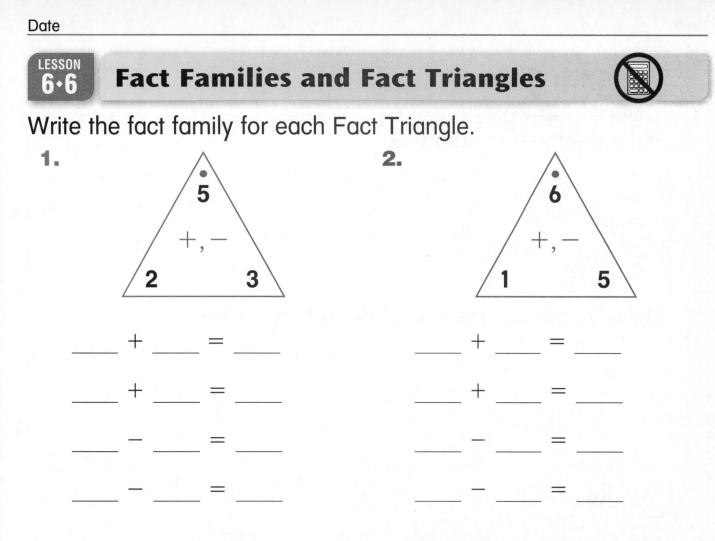

1.

△ 5 / +,− / 2   3

____ + ____ = ____

____ + ____ = ____

____ − ____ = ____

____ − ____ = ____

2.

△ 6 / +,− / 1   5

____ + ____ = ____

____ + ____ = ____

____ − ____ = ____

____ − ____ = ____

3.

△ 10 / +,− / 5   5

____ + ____ = ____

____ + ____ = ____

____ − ____ = ____

____ − ____ = ____

4. Write the missing number.
Write the fact family.

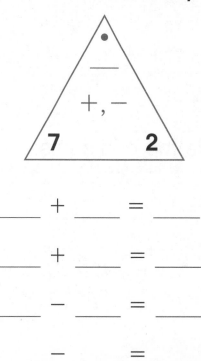

△ ___ / +,− / 7   2

____ + ____ = ____

____ + ____ = ____

____ − ____ = ____

____ − ____ = ____

Math Boxes

1. Write the missing numbers.

Rule
Count back
by 10s

| 44 | 34 | 24 | | |

2. Make a pattern. Use your Pattern-Block Template.

3. Find the sums.

6 + 1 = _____

_____ = 1 + 8

_____ = 2 + 2

0 + 4 = _____

Circle the odd sums.

4. **Number of Siblings**

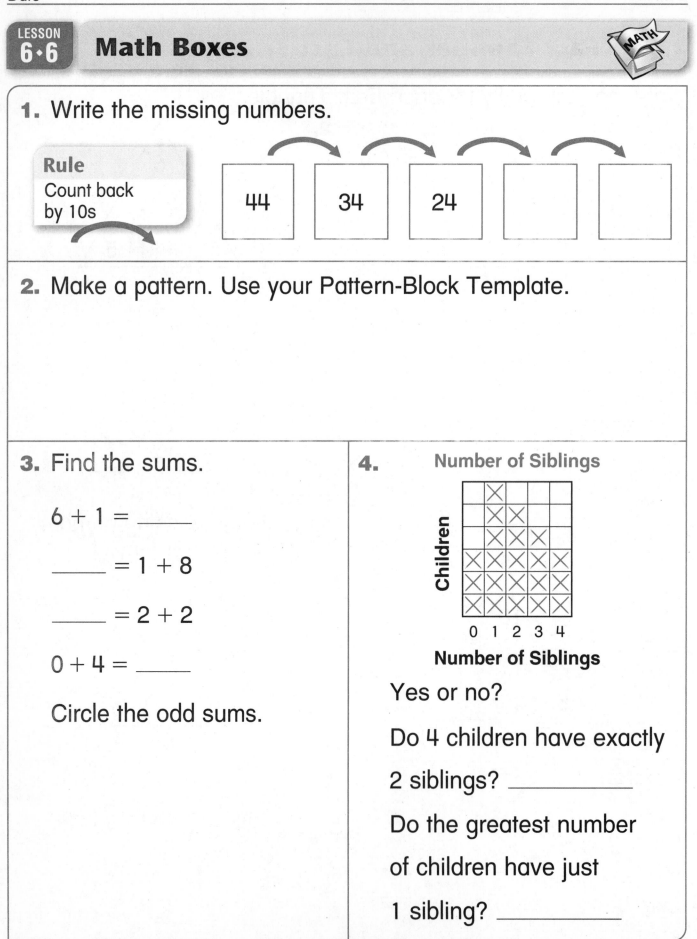

Children

0 1 2 3 4
Number of Siblings

Yes or no?

Do 4 children have exactly

2 siblings? _____

Do the greatest number

of children have just

1 sibling? _____

Date _____

1. Subtract.

$4 - 1 =$ _____

$7 - 2 =$ _____

$\begin{array}{r} 5 \\ -\ 4 \\ \hline \end{array}$
$\qquad$
$\begin{array}{r} 6 \\ -\ 0 \\ \hline \end{array}$

2. Write the fact family.

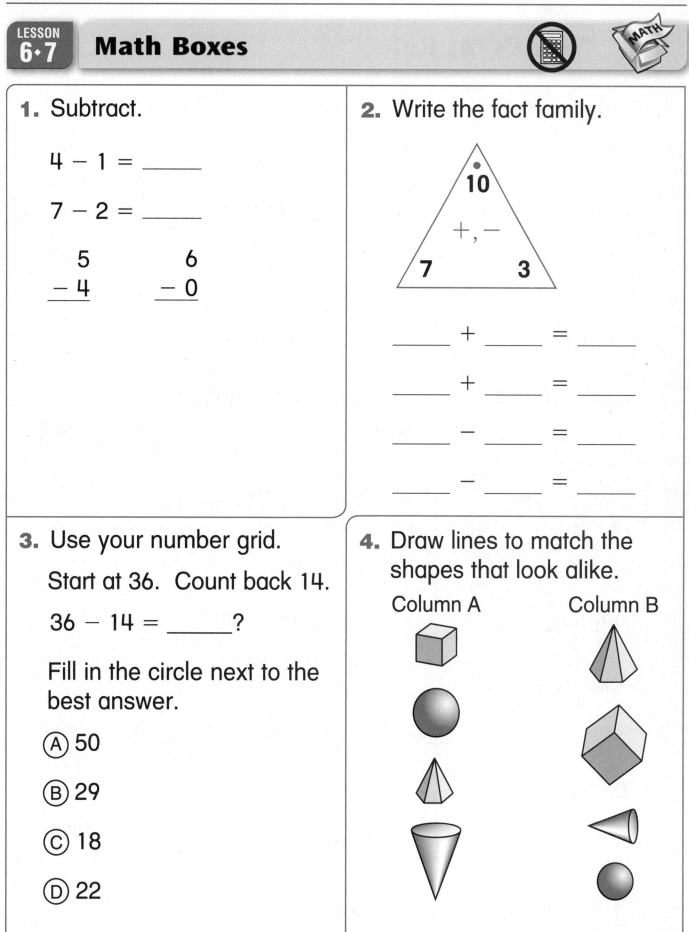

10

$+,-$

7   3

_____ + _____ = _____

_____ + _____ = _____

_____ − _____ = _____

_____ − _____ = _____

3. Use your number grid.

Start at 36. Count back 14.

$36 - 14 =$ _____?

Fill in the circle next to the best answer.

(A) 50

(B) 29

(C) 18

(D) 22

4. Draw lines to match the shapes that look alike.

Column A   Column B

LESSON 6·8

"What's My Rule?"

1. Find the rule.

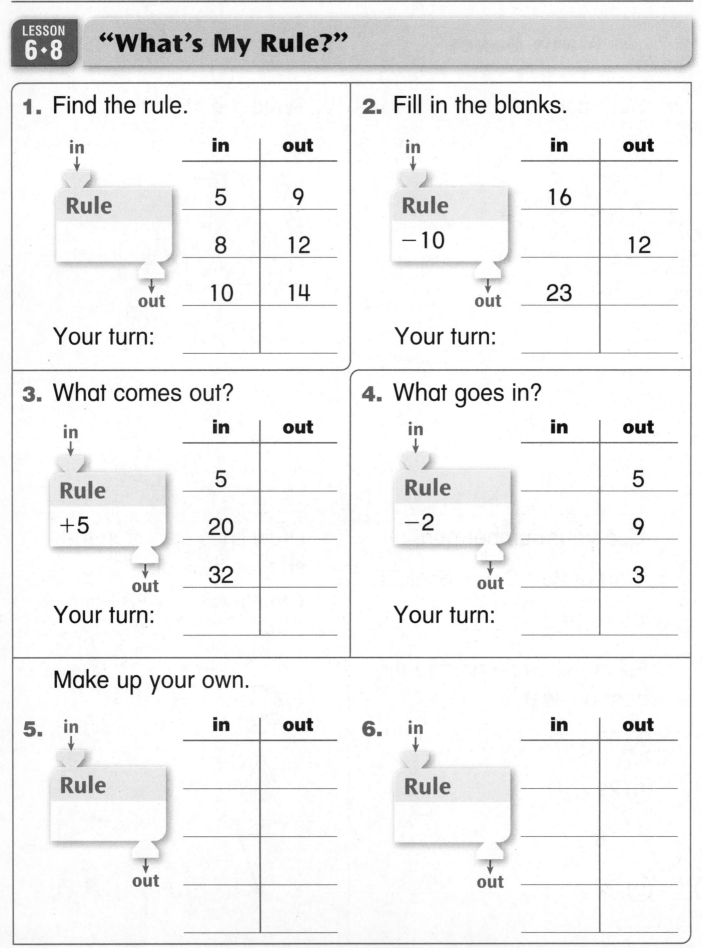

in → Rule [   ] → out

in	out
5	9
8	12
10	14

Your turn:

2. Fill in the blanks.

in → Rule −10 → out

in	out
16	
	12
23	

Your turn:

3. What comes out?

in → Rule +5 → out

in	out
5	
20	
32	

Your turn:

4. What goes in?

in → Rule −2 → out

in	out
	5
	9
	3

Your turn:

Make up your own.

5. in → Rule [   ] → out

in	out

6. in → Rule [   ] → out

in	out

LESSON 6·8

Math Boxes

1. Write the missing numbers.

Rule −2

32 [ ] 28 26 [ ]

2. Draw the next two figures.

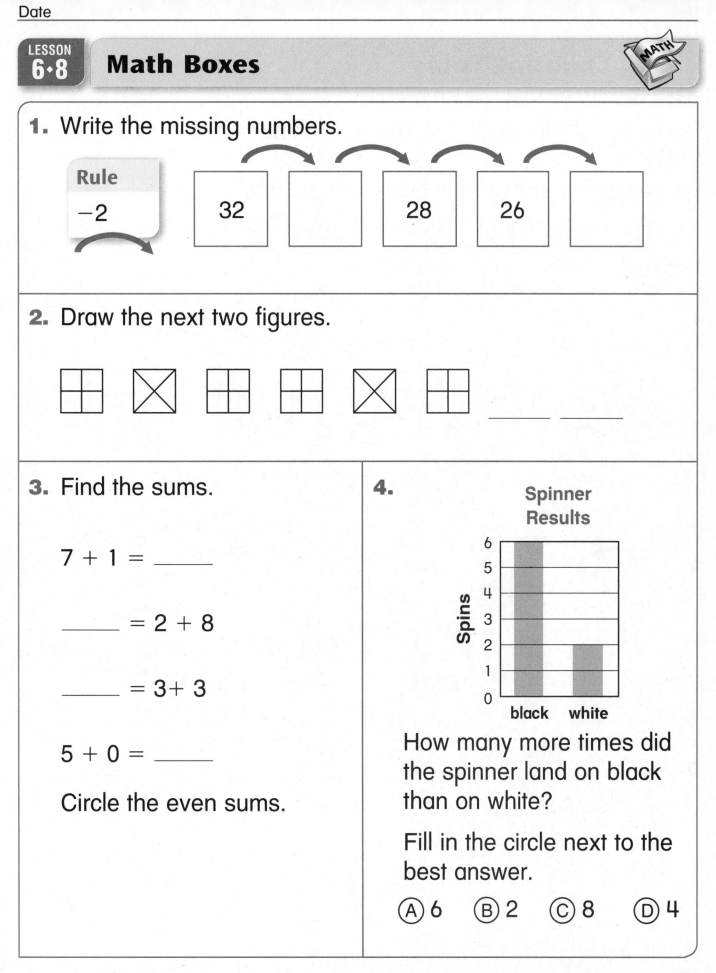

3. Find the sums.

$7 + 1 =$ _____

_____ $= 2 + 8$

_____ $= 3 + 3$

$5 + 0 =$ _____

Circle the even sums.

4.

Spinner Results

Spins

6
5
4
3
2
1
0

black   white

How many more times did the spinner land on black than on white?

Fill in the circle next to the best answer.

Ⓐ 6   Ⓑ 2   Ⓒ 8   Ⓓ 4

Date _____

P 1¢   N 5¢   D 10¢   Q 25¢
$0.01 $0.05 $0.10 $0.25
a penny a nickel a dime   a quarter

How much money? Use your coins.

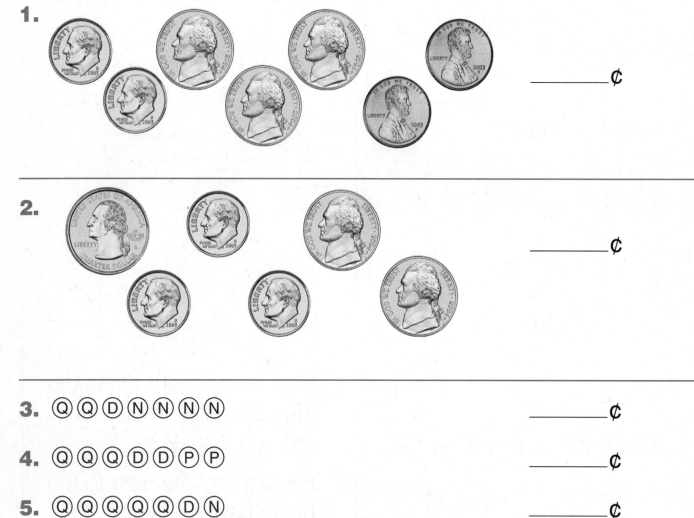

1. _____ ¢

2. _____ ¢

3. Q Q D N N N N   _____ ¢

4. Q Q Q D D P P   _____ ¢

5. Q Q Q Q Q D N   _____ ¢

LESSON 6·9 Math Boxes

1. Are you more likely to spin black or white?

_____

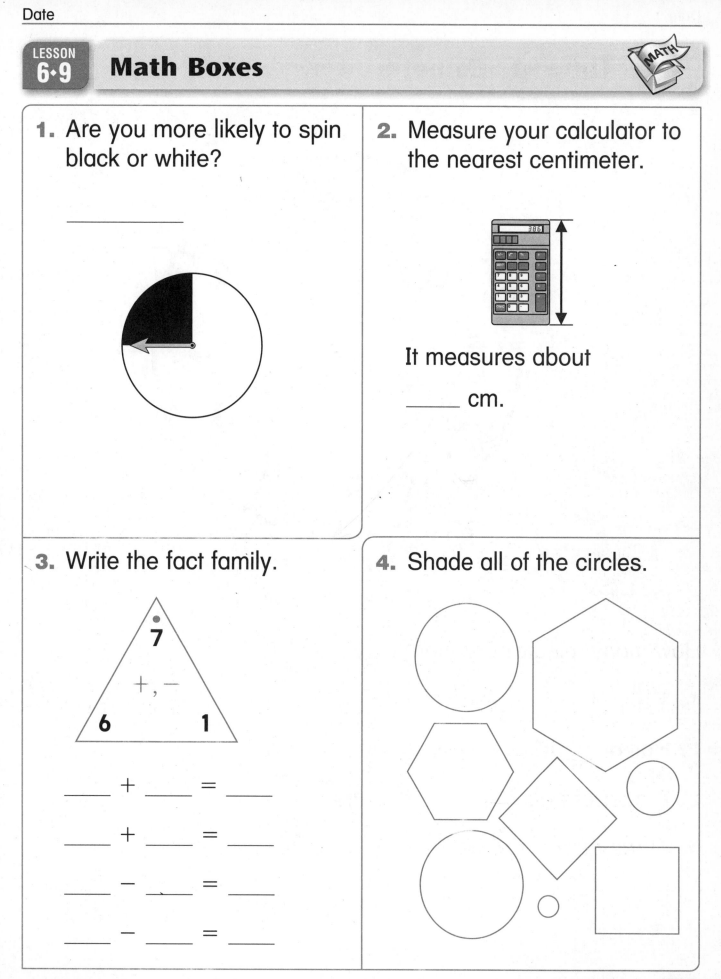

2. Measure your calculator to the nearest centimeter.

It measures about

_____ cm.

3. Write the fact family.

7

+, −

6   1

____ + ____ = ____

____ + ____ = ____

____ − ____ = ____

____ − ____ = ____

4. Shade all of the circles.

LESSON 6·10   Time at 5-Minute Intervals

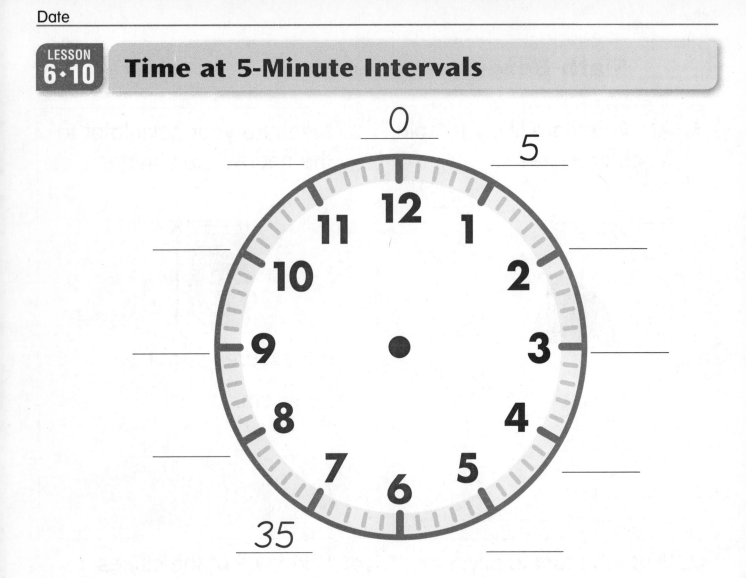

How many minutes are there in:

1. 1 hour? _____ minutes

2. Half an hour? _____ minutes

3. A quarter-hour? _____ minutes

4. Three-quarters of an hour? _____ minutes

LESSON 6·10 **Digital Notation**

Draw the hour hand and the minute hand.

1.

4:00

2.

2:30

3.

6:15

Write the time.

4.

____ : ____

5.

____ : ____

6.

____ : ____

Make up your own. Draw the hour hand and minute hand.
Write the time.

7.

____ : ____

8.

____ : ____

9.

____ : ____

LESSON 6·10 Math Boxes

1. Measure your shoe.

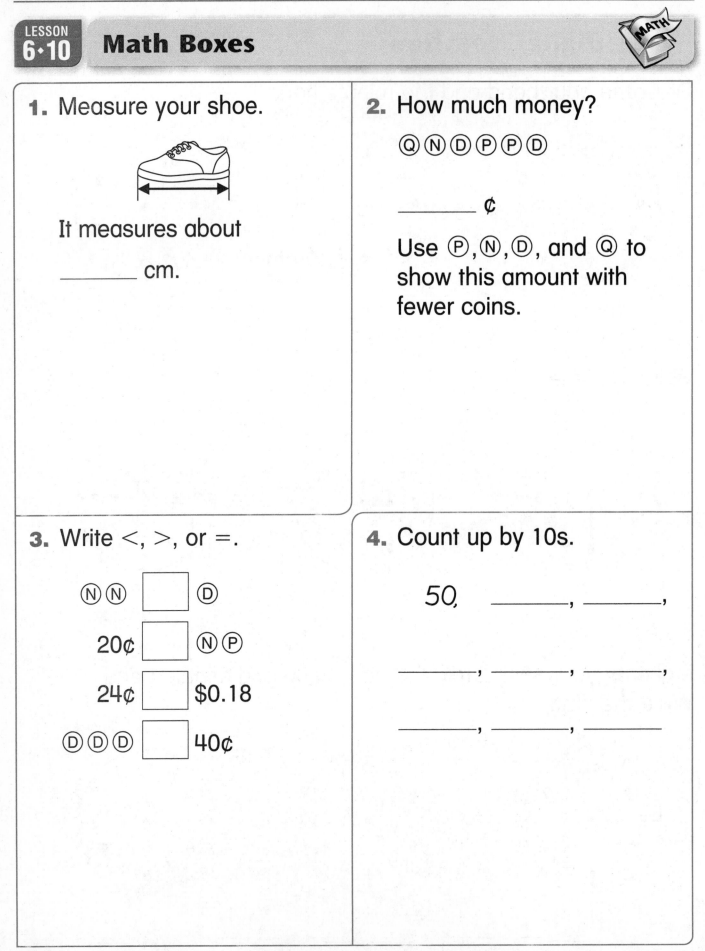

It measures about

_____ cm.

2. How much money?

Ⓠ Ⓝ Ⓓ Ⓟ Ⓟ Ⓓ

_____ ¢

Use Ⓟ, Ⓝ, Ⓓ, and Ⓠ to show this amount with fewer coins.

3. Write <, >, or =.

Ⓝ Ⓝ ☐ Ⓓ

20¢ ☐ Ⓝ Ⓟ

24¢ ☐ $0.18

Ⓓ Ⓓ Ⓓ ☐ 40¢

4. Count up by 10s.

50, _____, _____,

_____, _____, _____,

_____, _____, _____

Date _____

My Reference Book Scavenger Hunt

START
Turn to the Table of Contents.

Which section is about Measurement?
Fill in the circle next to the best answer.

Ⓐ 5th section   Ⓑ 3rd section

Ⓒ 4th section   Ⓓ 6th section

Find 2 tools in the measurement section. Draw them.

This is on page _____. │ This is on page _____.

Turn to page 96.

Draw one pattern that you see.

Turn to the Games Section.

Find your favorite first-grade math game.

My favorite first-grade math game is

_____.

My favorite first-grade math game is on page

_____.

END

LESSON 6·11 1- and 10-Centimeter Objects

1. Find 3 things that are about 1 centimeter long.

   Use words or pictures to show the things you found.

2. Find 3 things that are about 10 centimeters long.

   Use words or pictures to show the things you found.

Date _____

1. Are you more likely to spin black or white?

_____

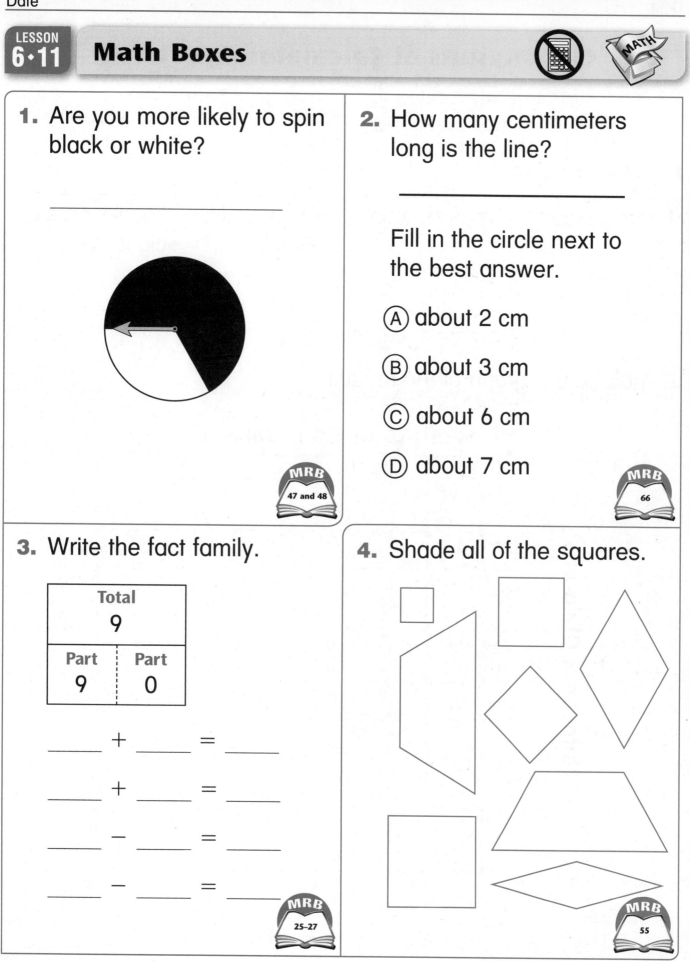

MRB
47 and 48

2. How many centimeters long is the line?

_____

Fill in the circle next to the best answer.

(A) about 2 cm

(B) about 3 cm

(C) about 6 cm

(D) about 7 cm

MRB
66

3. Write the fact family.

Total	
9	
Part	**Part**
9	0

____ + ____ = ____

____ + ____ = ____

____ − ____ = ____

____ − ____ = ____

MRB
25–27

4. Shade all of the squares.

MRB
55

Date _____

Class Results of Calculator Counts

1. I counted to _____ in 15 seconds.

2. Class results:

Largest count	Smallest count	Range of class counts	Middle value of class counts
_____	_____	_____	_____

3. Make a bar graph of the results.

Results of Calculator Counts

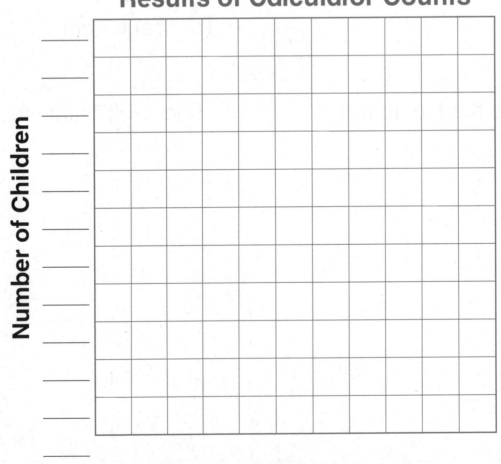

Number of Children

Counted to

Math Boxes

1. Draw a line segment that is about 7 centimeters long.

MRB
66

2. How much money is
Ⓠ Ⓠ Ⓠ Ⓟ Ⓟ Ⓟ?

Fill in the circle next to the best answer.

Ⓐ 78¢

Ⓑ 33¢

Ⓒ 73¢

Ⓓ 45¢

MRB
88 and 89

3. Write <, >, or =.

7 + 6 ☐ 12

13 ☐ 6 + 7

14 − 6 ☐ 7

8 ☐ 15 − 6

MRB
9

4. Count up by 5s.

25, —————, —————,

—————, —————, —————,

—————, —————, —————

Math Boxes

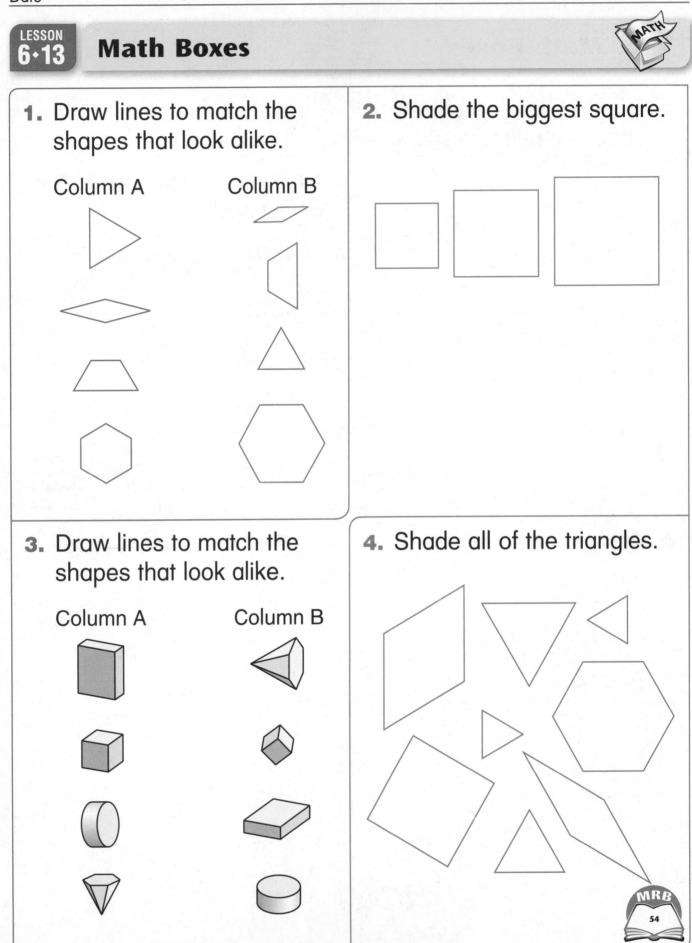

1. Draw lines to match the shapes that look alike.

Column A   Column B

2. Shade the biggest square.

3. Draw lines to match the shapes that look alike.

Column A   Column B

4. Shade all of the triangles.

MRB
54

LESSON 7·1

Make My Design

Materials
- ☐ pattern blocks
- ☐ folder

Players 2

Skill Create designs using pattern blocks

Object of the Game To create a design identical to the other player's design

Directions:

1. The first player chooses 6 blocks. The second player gathers the same 6 blocks.

2. Players sit face-to-face with a folder between them.

3. The first player creates a design with the blocks.

4. Using only words, the first player tells the second player how to "Make My Design." The second player can ask questions about the instructions.

5. Players remove the folder and look at the two designs. Players discuss how closely the designs match.

6. Players change roles and play again.

LESSON 7·1 Math Boxes

1. Shade the large shapes.

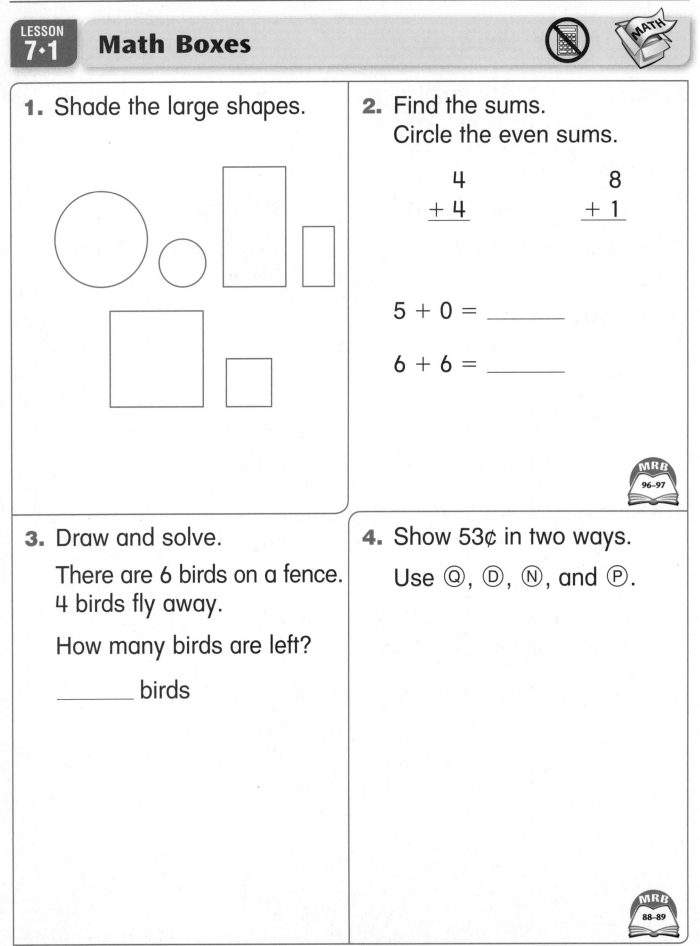

2. Find the sums.
Circle the even sums.

$$\begin{array}{r} 4 \\ +\ 4 \\ \hline \end{array} \qquad \begin{array}{r} 8 \\ +\ 1 \\ \hline \end{array}$$

$5 + 0 =$ _____

$6 + 6 =$ _____

MRB
96–97

3. Draw and solve.

There are 6 birds on a fence. 4 birds fly away.

How many birds are left?

_____ birds

4. Show 53¢ in two ways.

Use Ⓠ, Ⓓ, Ⓝ, and Ⓟ.

MRB
88–89

LESSON 7·2 Math Boxes

1. Draw what comes next.

☐   ☐☐   ☐☐☐   ☐☐☐☐   _____

2. Subtract.

$5 - 1 =$ _____   _____ $= 4 - 0$

$$\begin{array}{r} 3 \\ -\ 3 \\ \hline \end{array}$$ $$\begin{array}{r} 6 \\ -\ 1 \\ \hline \end{array}$$

3. Draw the hands.

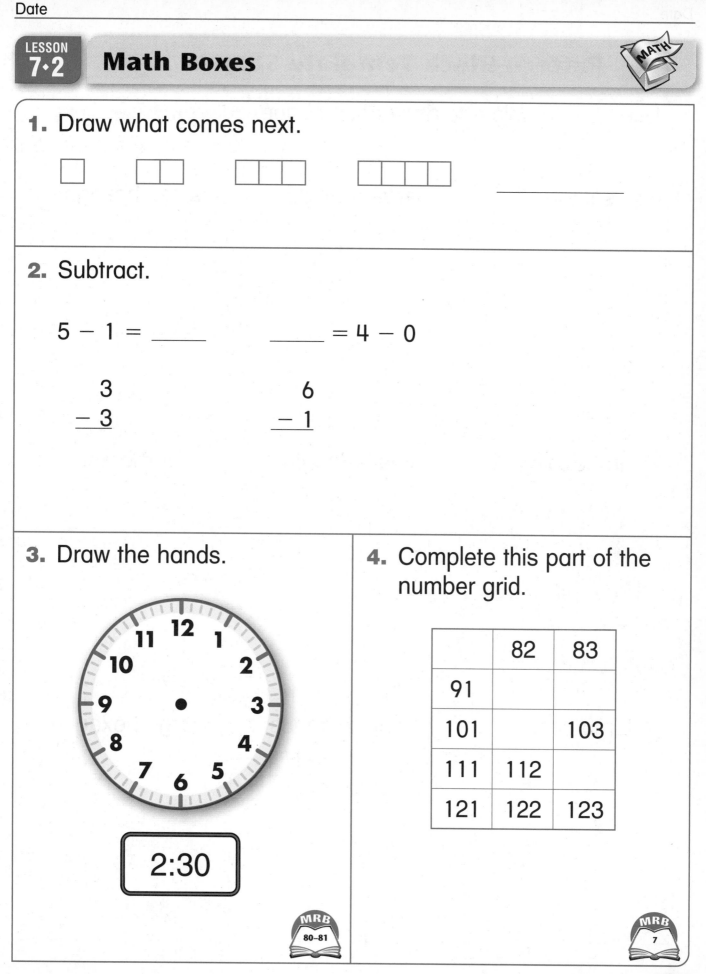

2:30

4. Complete this part of the number grid.

	82	83
91		
101		103
111	112	
121	122	123

MRB 80–81

MRB 7

Date _____

Pattern-Block Template Shapes

1. Use your template to draw each shape.

square	large triangle	small hexagon
trapezoid	small triangle	fat rhombus
large circle	skinny rhombus	large hexagon

Date

LESSON 7•3

Pattern-Block Template Shapes *continued*

2. Draw shapes that have exactly 4 sides and 4 corners. Write their names.

LESSON 7·3 Math Boxes

1. Find the small square.
Shade it.

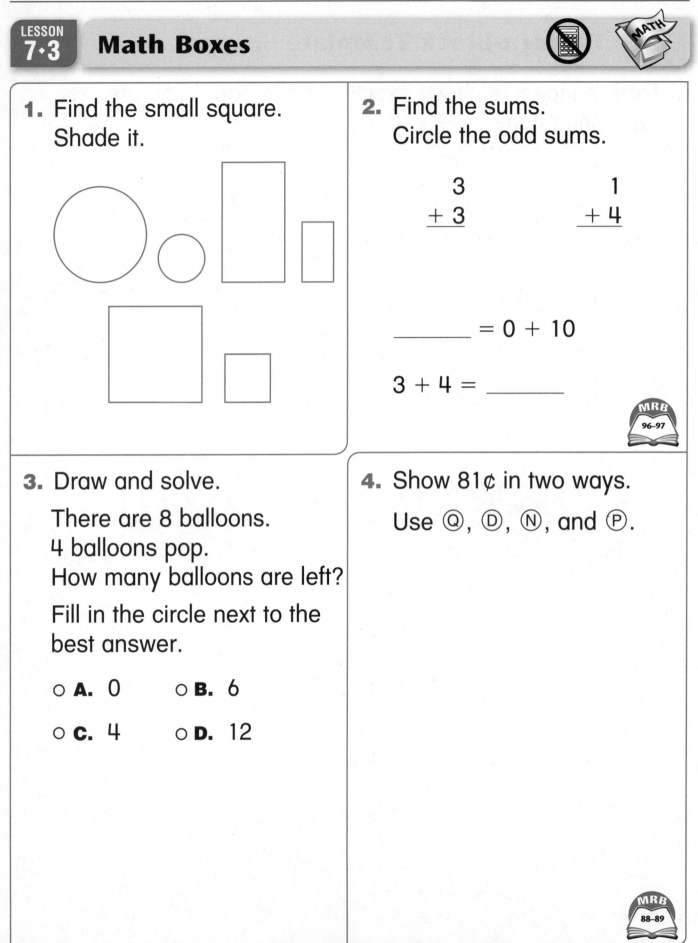

2. Find the sums.
Circle the odd sums.

$$\begin{array}{r} 3 \\ + 3 \\ \hline \end{array} \qquad \begin{array}{r} 1 \\ + 4 \\ \hline \end{array}$$

_____ = 0 + 10

3 + 4 = _____

MRB
96–97

3. Draw and solve.

There are 8 balloons.
4 balloons pop.
How many balloons are left?

Fill in the circle next to the best answer.

○ **A.** 0 ○ **B.** 6

○ **C.** 4 ○ **D.** 12

4. Show 81¢ in two ways.

Use ⓠ, ⓓ, ⓝ, and ⓟ.

MRB
88–89

Date

Triangles

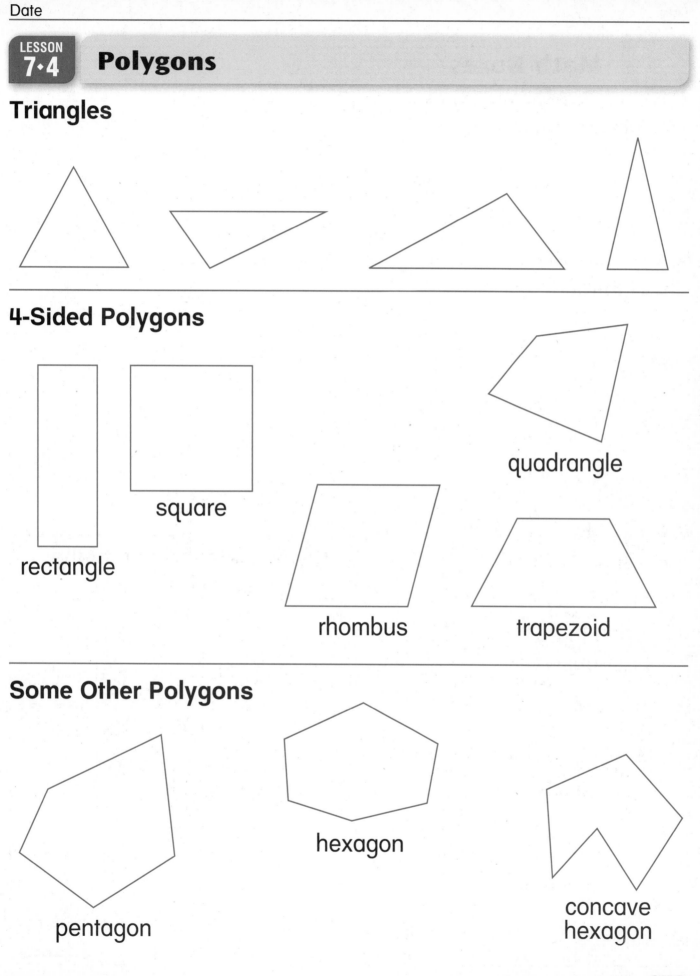

4-Sided Polygons

rectangle

square

quadrangle

rhombus

trapezoid

Some Other Polygons

pentagon

hexagon

concave
hexagon

LESSON 7·4 Math Boxes

1. Draw what comes next.

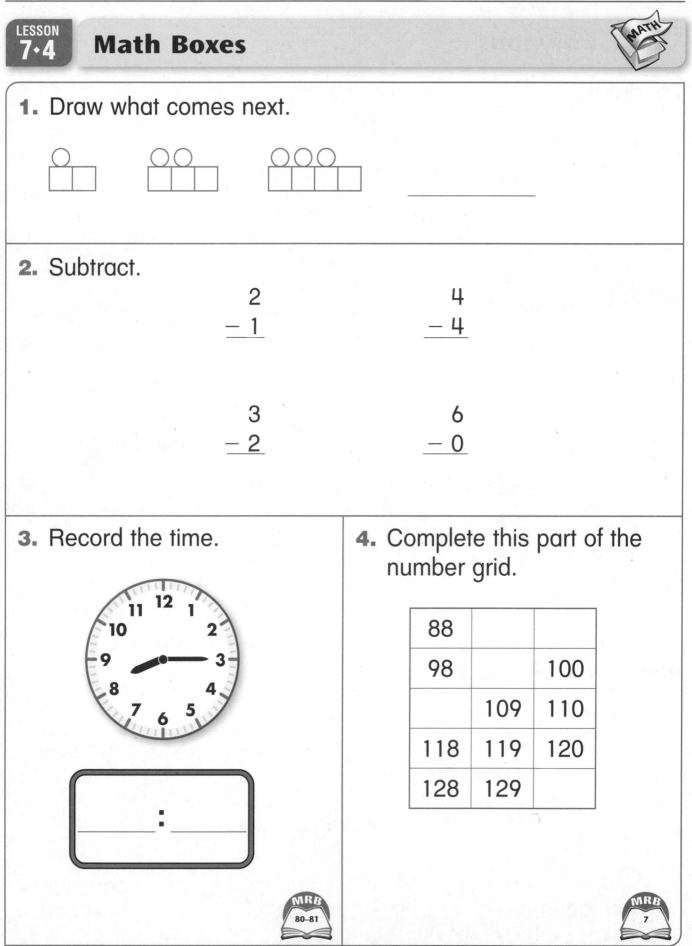

2. Subtract.

$$2 - 1$$ $$4 - 4$$

$$3 - 2$$ $$6 - 0$$

3. Record the time.

4. Complete this part of the number grid.

88		
98		100
	109	110
118	119	120
128	129	

MRB 80–81

MRB 7

Date _____

1. Circle the name of this 3-dimensional shape.

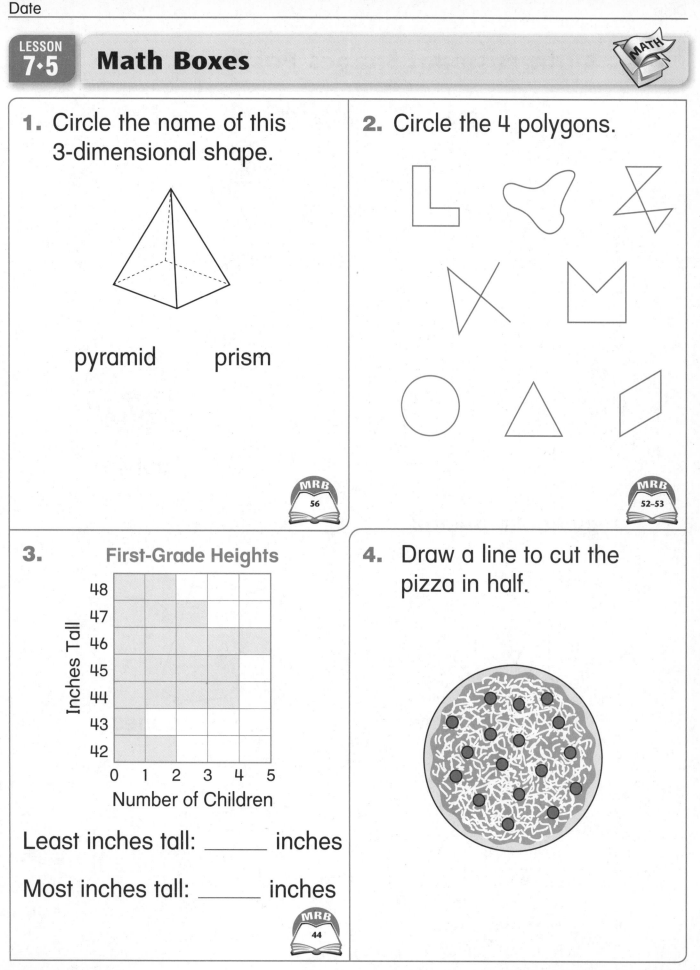

pyramid   prism

MRB 56

2. Circle the 4 polygons.

MRB 52–53

3. First-Grade Heights

Inches Tall

48
47
46
45
44
43
42

0 1 2 3 4 5
Number of Children

Least inches tall: _____ inches

Most inches tall: _____ inches

MRB 44

4. Draw a line to cut the pizza in half.

LESSON 7·6 3-Dimensional Shapes Poster

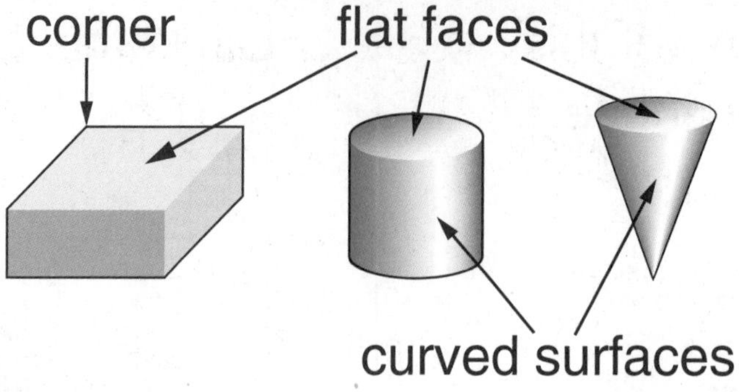

corner   flat faces

curved surfaces

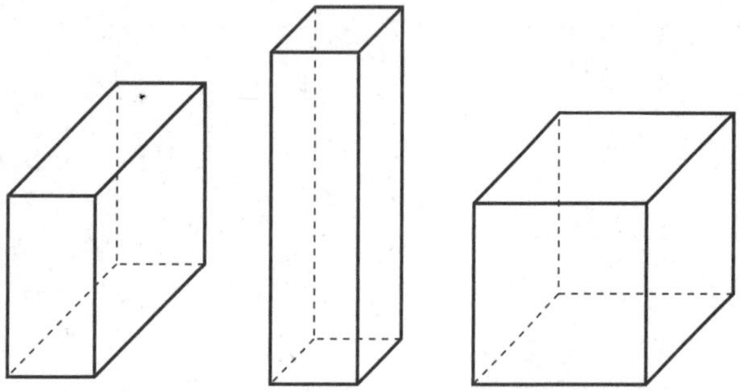

rectangular prisms

cube

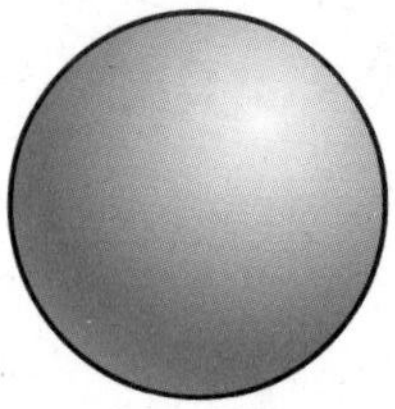

sphere

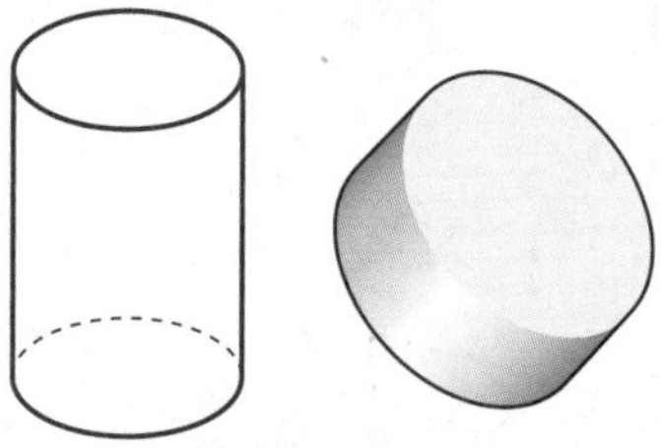

cylinders

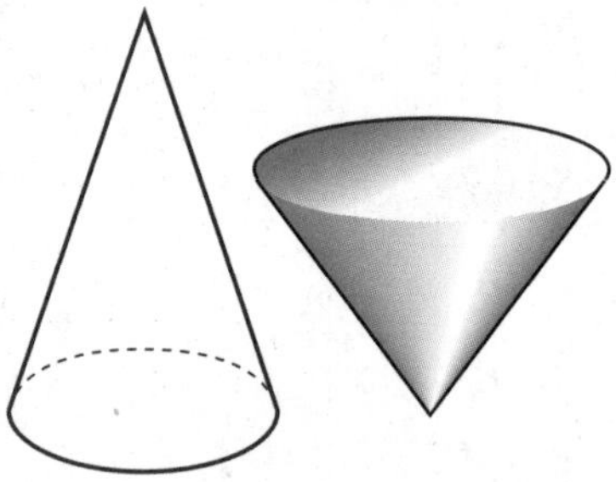

cones

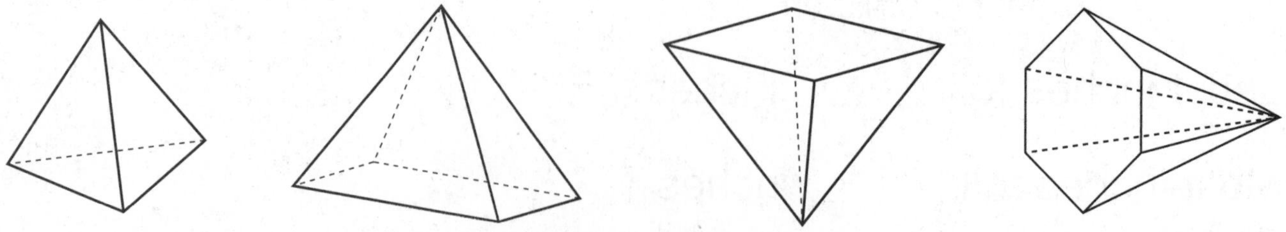

pyramids

Date _____

What kind of shape is each object?
Write its name under the picture.

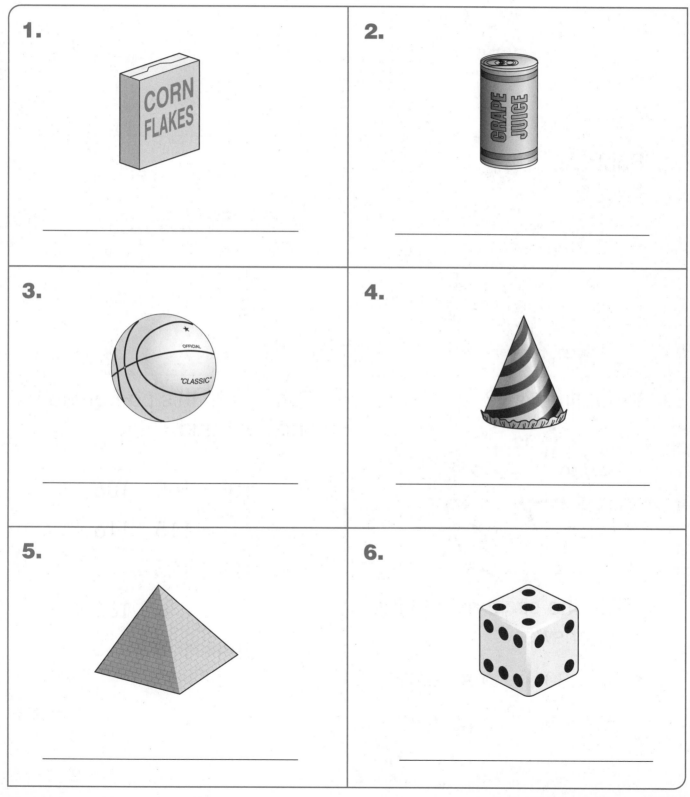

1. _____

2. _____

3. _____

4. _____

5. _____

6. _____

LESSON 7·6 Math Boxes

1. Draw what comes next.

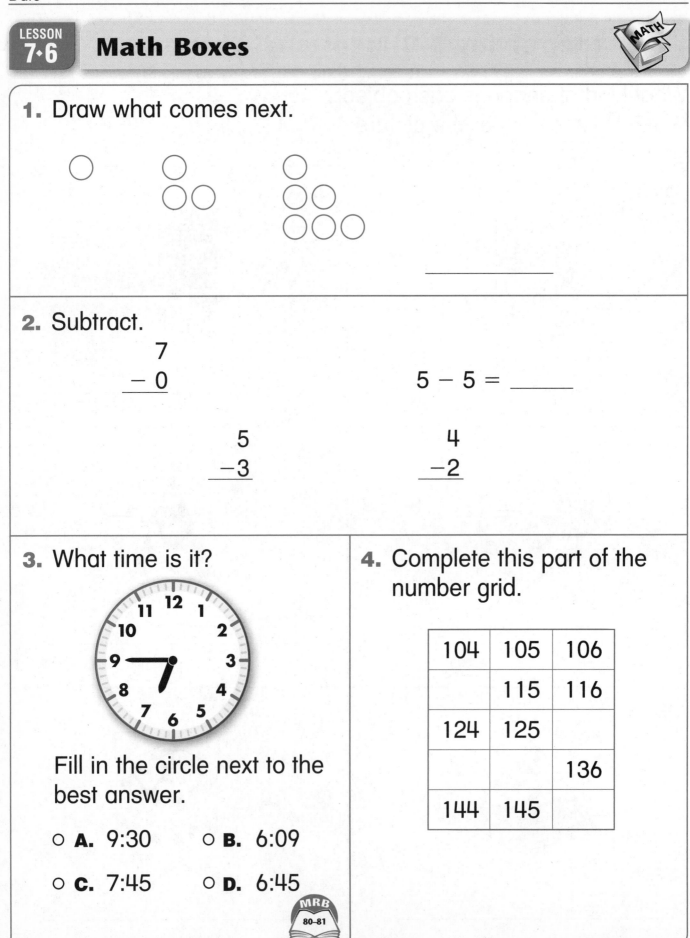

_____

2. Subtract.

$$7 - 0$$

$$5 - 5 = \underline{\qquad}$$

$$5 - 3$$

$$4 - 2$$

3. What time is it?

Fill in the circle next to the best answer.

○ **A.** 9:30   ○ **B.** 6:09

○ **C.** 7:45   ○ **D.** 6:45

4. Complete this part of the number grid.

104	105	106
	115	116
124	125	
		136
144	145	

MRB 80–81

LESSON 7·7 **Math Boxes**

1. Name or draw 3 cylinders in your classroom.

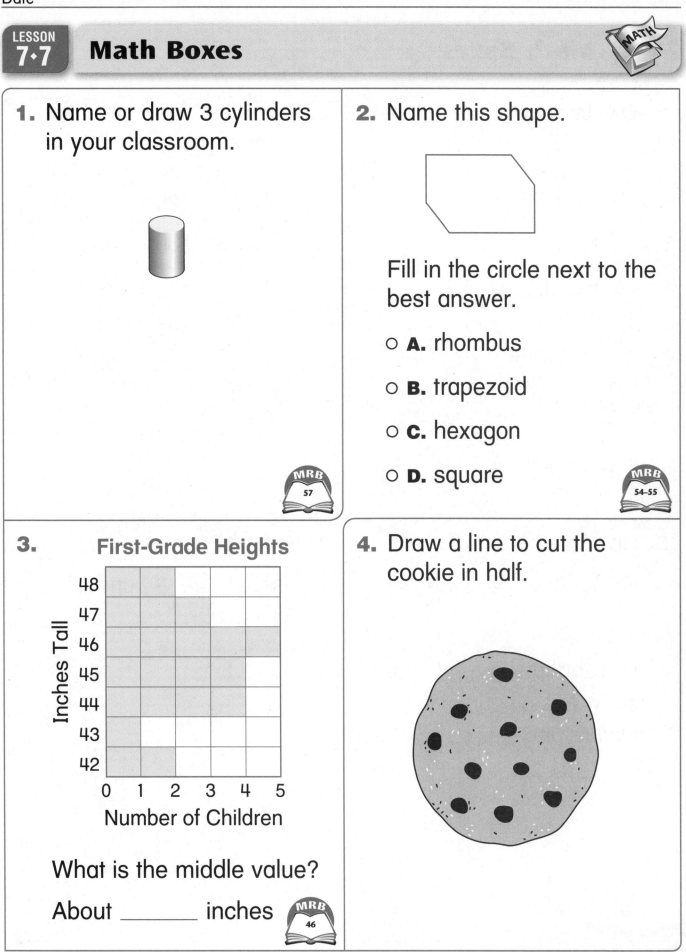

2. Name this shape.

Fill in the circle next to the best answer.

○ **A.** rhombus

○ **B.** trapezoid

○ **C.** hexagon

○ **D.** square

MRB
54–55

MRB
57

3. **First-Grade Heights**

Inches Tall

48
47
46
45
44
43
42

0 1 2 3 4 5
Number of Children

What is the middle value?

About _____ inches

MRB
46

4. Draw a line to cut the cookie in half.

LESSON 7·8 Math Boxes

1. Divide each shape in half.

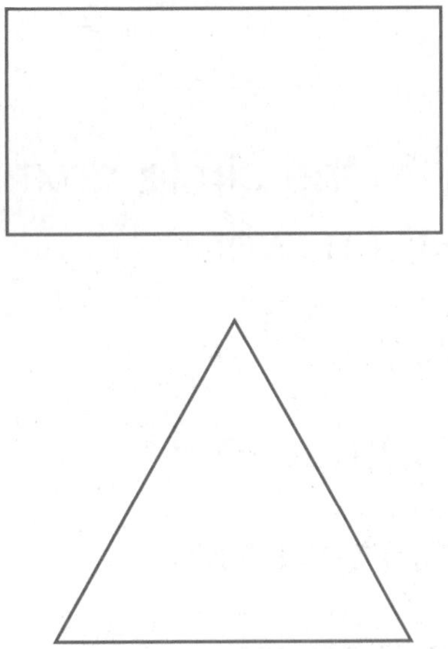

2. Complete this part of the number grid.

123	124	
133	134	135
		145
153	154	
	164	165

3. How much money?

Ⓠ Ⓝ Ⓓ Ⓝ Ⓓ Ⓟ Ⓝ

_____ ¢

Use Ⓠ, Ⓓ, Ⓝ, and Ⓟ to show this amount with fewer coins.

4. Show 65¢ in two ways.

Use Ⓠ, Ⓓ, Ⓝ, and Ⓟ.

MRB 88–89

MRB 88–89

Date _____

How Much Money?

Record the amount shown.

1.
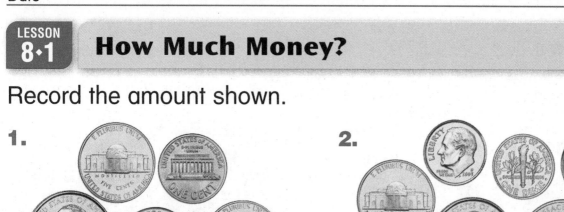

2.

_____ ¢   _____ ¢

Mark the coins you need to buy each item.

3.

86¢
crystal

4.

59¢
horse

Time

Draw the hands.

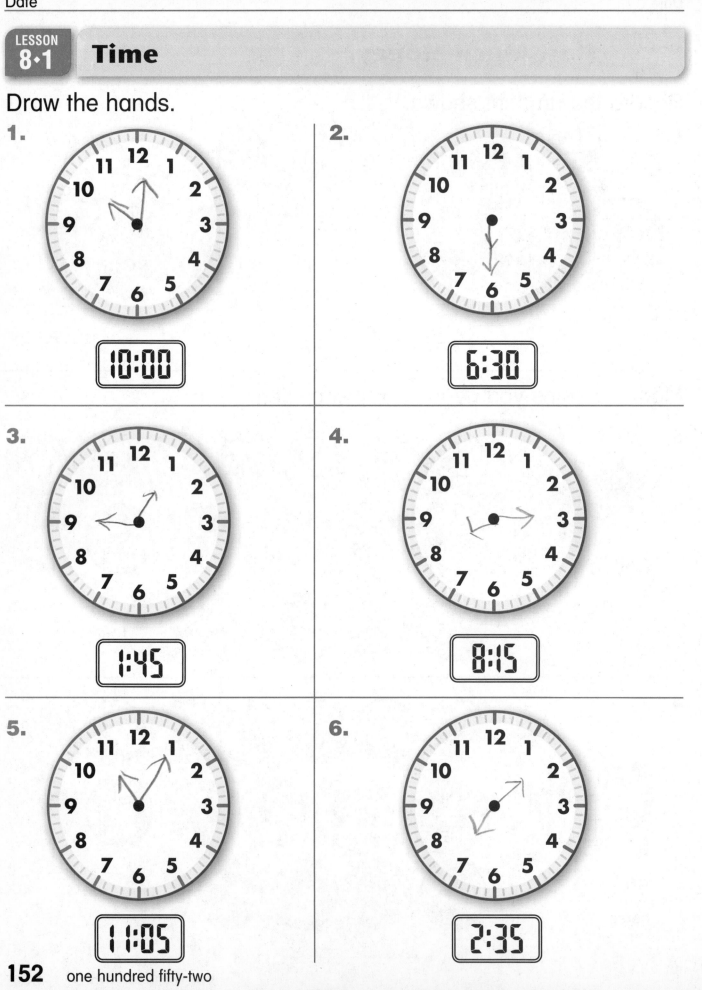

1.

10:00

2.

6:30

3.

1:45

4.

8:15

5.

11:05

6.

2:35

Date

Math Boxes

1. How much money?

Q Q Q Q Q D P

_____ ¢ or

$ _____

2. Draw a line to match each face to the correct picture of the 3-dimensional shape.

Column A   Column B

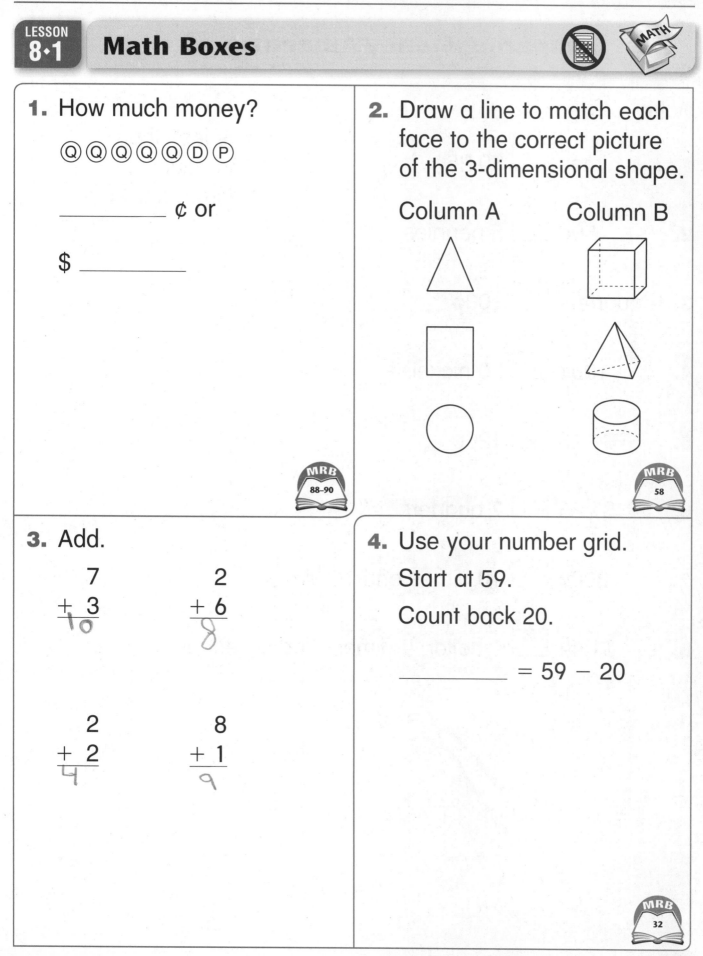

MRB
58

3. Add.

```
  7        2
+ 3      + 6
—10       —8
```

```
  2        8
+ 2      + 1
—4        —9
```

4. Use your number grid.

Start at 59.

Count back 20.

_____ = 59 − 20

MRB
32

LESSON 8·2   Comparing Money Amounts

Write <, >, or =.

> < is less than
> = is equal to
> > is greater than

1. 2 dimes ☐ $0.25

2. 50¢ ☐ 5 pennies

3. 4 quarters ☐ 100¢

4. 100¢ ☐ 20 nickels

5. $1.25 ☐ 120¢

6. $1.75 ☐ 7 quarters

7. 200¢ ☐ 10 dimes and 10 nickels

8. $1.44 ☐ 1 dollar, 4 dimes, and 4 pennies

LESSON 8·2 Math Boxes

1. Use $1, Q, D, N, and P to show this amount.

$3.49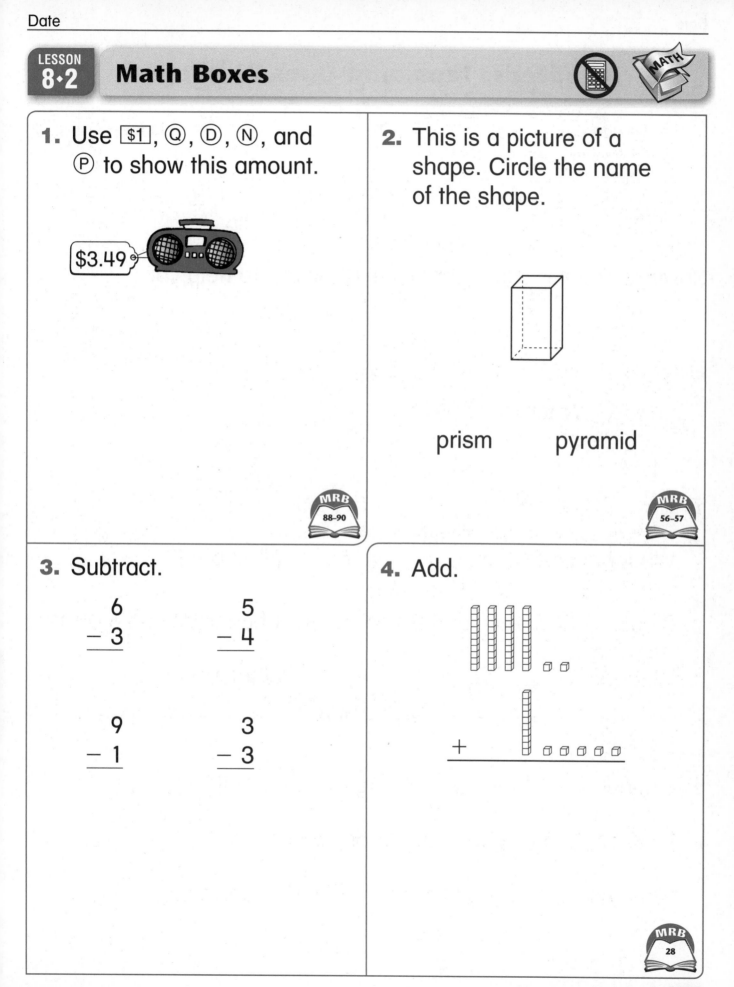

MRB 88–90

2. This is a picture of a shape. Circle the name of the shape.

prism   pyramid

MRB 56–57

3. Subtract.

$$6 - 3$$ $$5 - 4$$

$$9 - 1$$ $$3 - 3$$

4. Add.

MRB 28

Date _____

Hundreds, Tens, and Ones Riddles

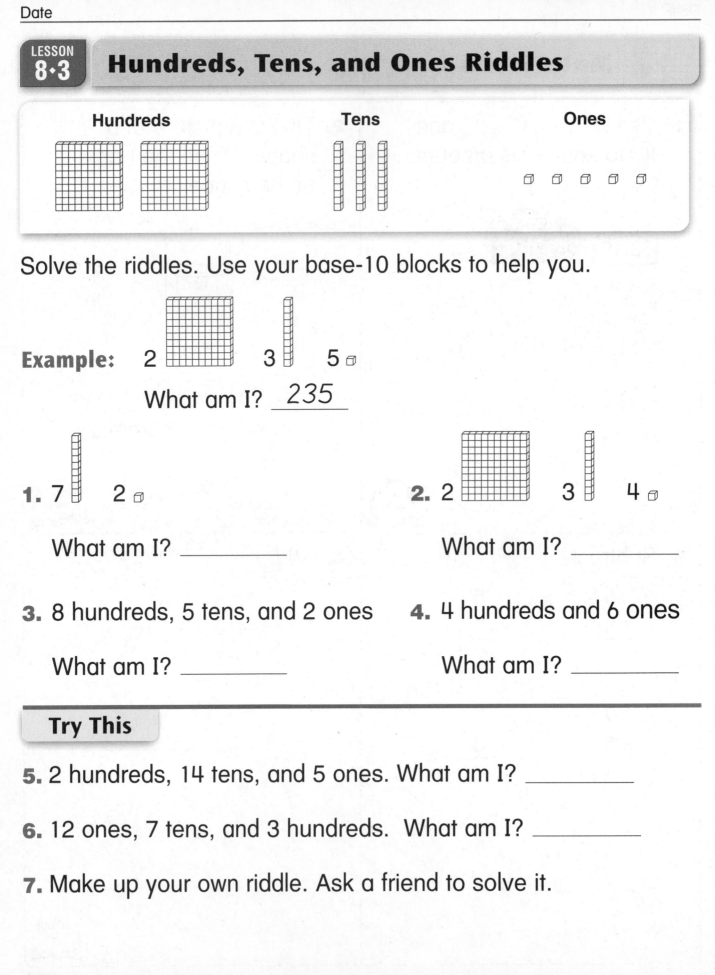

Hundreds	Tens	Ones

Solve the riddles. Use your base-10 blocks to help you.

Example: 2 ▦ 3 ▯ 5 ▫

What am I? ___235___

1. 7 ▯ 2 ▫

What am I? _____

2. 2 ▦ 3 ▯ 4 ▫

What am I? _____

3. 8 hundreds, 5 tens, and 2 ones

What am I? _____

4. 4 hundreds and 6 ones

What am I? _____

Try This

5. 2 hundreds, 14 tens, and 5 ones. What am I? _____

6. 12 ones, 7 tens, and 3 hundreds. What am I? _____

7. Make up your own riddle. Ask a friend to solve it.

LESSON 8·3 Math Boxes

1. Count the coins.

Q Q Q N P P P

Choose the best answer.

(A) 38¢

(B) 88¢

(C) 93¢

(D) 83¢

MRB 88–90

2. Draw a line to match each face to the correct picture of the 3-dimensional shape.

Column A   Column B

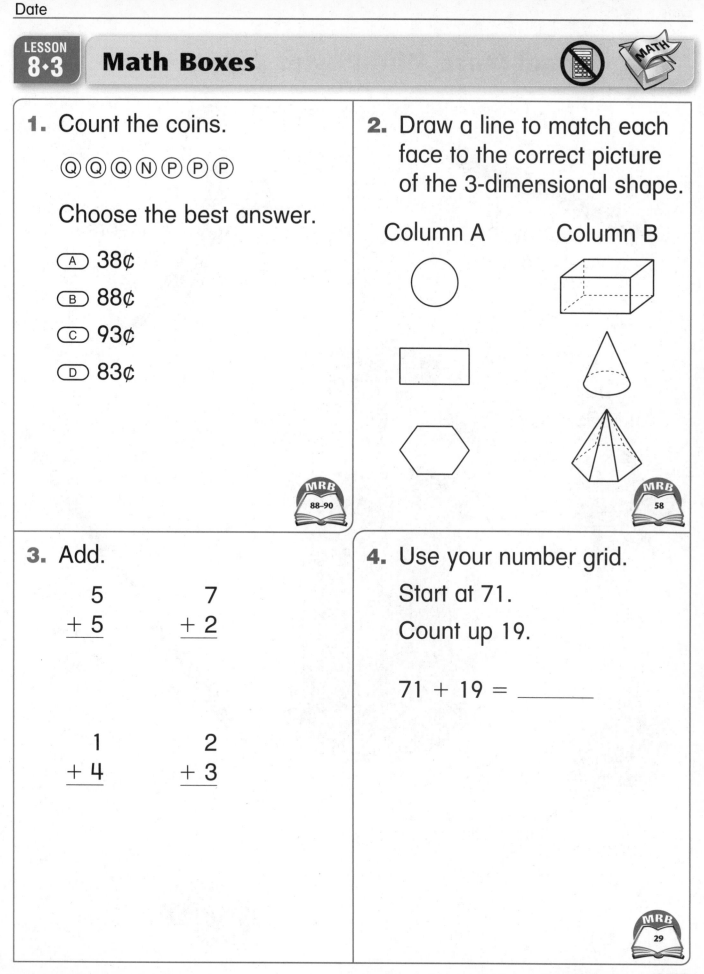

MRB 58

3. Add.

```
  5          7
+ 5        + 2
```

```
  1          2
+ 4        + 3
```

4. Use your number grid.

Start at 71.

Count up 19.

71 + 19 = _____

MRB 29

Date

School Store Mini-Poster 2

crayon
6¢

scissors
32¢

ball
35¢

gum
2¢

pencil
28¢

candy
8¢

eraser
17¢

LESSON 8·4

School Store Mini-Poster 3

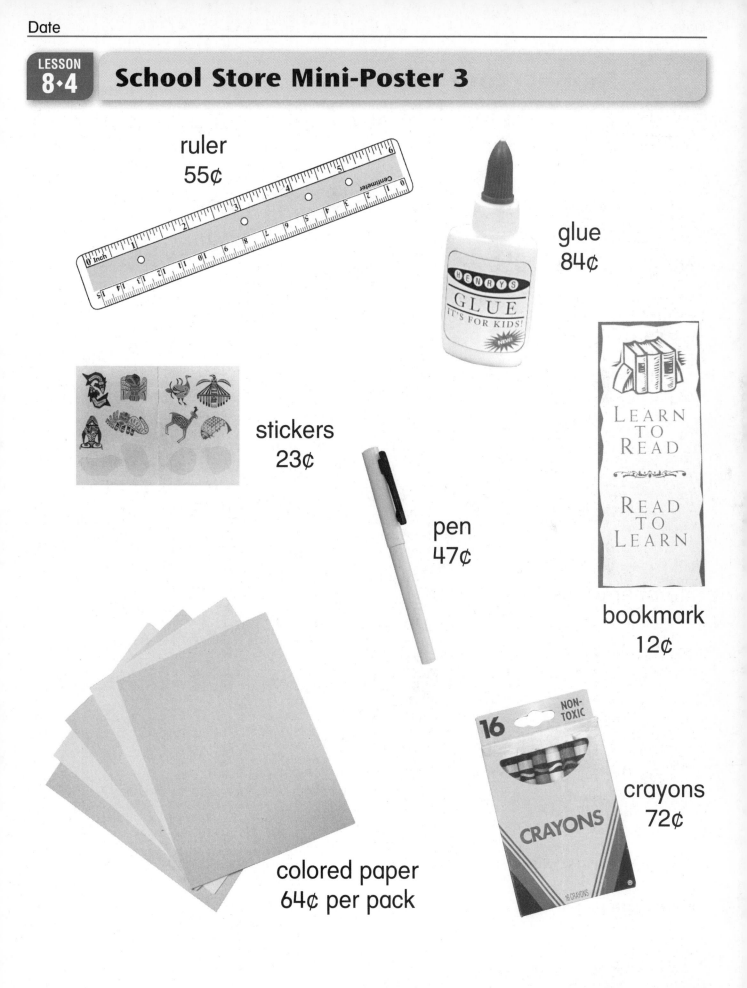

ruler
55¢

glue
84¢

stickers
23¢

pen
47¢

bookmark
12¢

colored paper
64¢ per pack

crayons
72¢

LESSON 8·4 Number Stories

Sample Story

I bought a ⚾ and an ✏. I paid 52 cents.

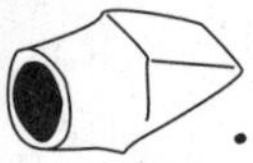

Number model: 35¢ + 17¢ = 52¢

1. Story 1

Number model: _____

2. Story 2

Number model: _____

Date

Math Boxes

1. $1.00 =

_____ pennies

_____ nickels

_____ dimes

_____ quarters

MRB
88–90

2. Name or draw 2 objects shaped like a rectangular prism.

MRB
56–57

3. Subtract.

$$\begin{array}{r} 4 \\ -2 \\ \hline \end{array} \qquad \begin{array}{r} 7 \\ -1 \\ \hline \end{array}$$

$$\begin{array}{r} 6 \\ -0 \\ \hline \end{array} \qquad \begin{array}{r} 10 \\ -5 \\ \hline \end{array}$$

4. What is the sum?

Choose the best answer.

Ⓐ 45   Ⓑ 87

Ⓒ 85   Ⓓ 78

MRB
28

LESSON 8·5 Museum Store Mini-Poster

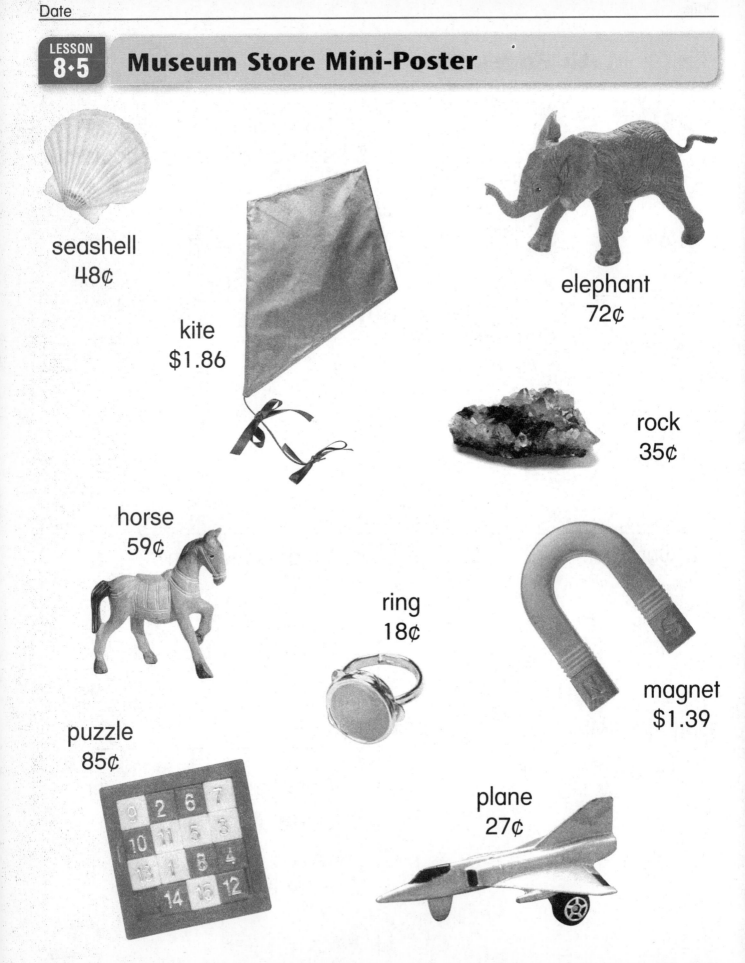

seashell
48¢

kite
$1.86

elephant
72¢

rock
35¢

horse
59¢

ring
18¢

magnet
$1.39

puzzle
85¢

plane
27¢

LESSON 8·5

Making Change

Record what you bought. Record how much change you got.

Example:

I bought _____*a plane*_____ for ___27___ cents.

I gave _____D D D_____
to the clerk.

I got _____P P P_____ in change.

---

1. I bought _____ for _____ cents.

I gave _____ to the clerk.

I got _____ in change.

---

2. I bought _____ for _____ cents.

I gave _____ to the clerk.

I got _____ in change.

---

3. I bought _____ for _____ cents.

I gave _____ to the clerk.

I got _____ in change.

Date

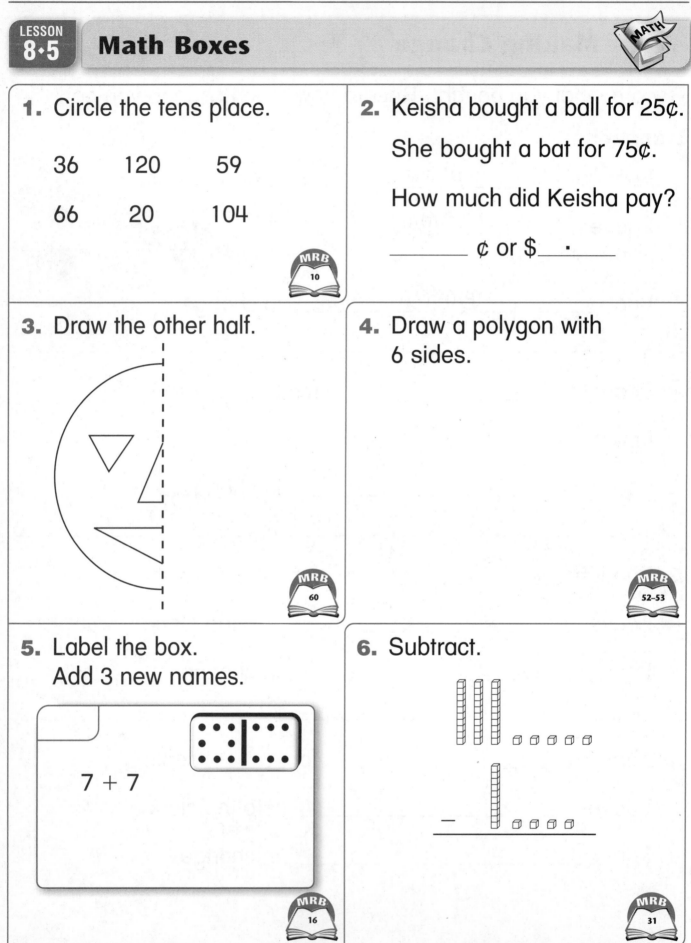

1. Circle the tens place.

36   120   59

66   20   104

MRB 10

2. Keisha bought a ball for 25¢.

She bought a bat for 75¢.

How much did Keisha pay?

_____ ¢ or $____ . ____

3. Draw the other half.

MRB 60

4. Draw a polygon with 6 sides.

MRB 52–53

5. Label the box.
Add 3 new names.

7 + 7

MRB 16

6. Subtract.

MRB 31

LESSON 8·6   Equal Shares

Show how you share your crackers.

1 cracker, 2 people	1 cracker, 4 people
Halves	**Fourths**

1 cracker, 3 people	2 crackers, 4 people
Thirds	

LESSON 8·6 "What's My Rule?"

Write the rule. Complete the table.

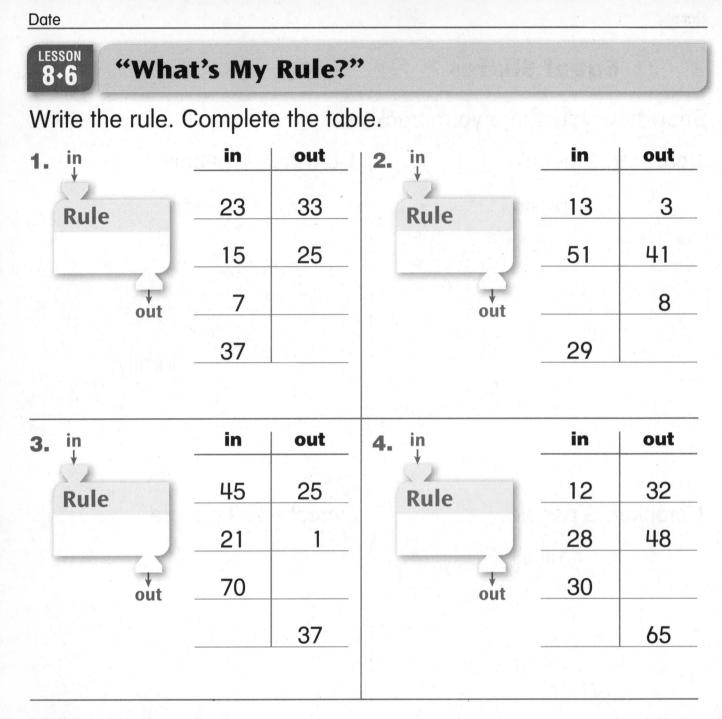

1. in → Rule → out

in	out
23	33
15	25
7	
37	

2. in → Rule → out

in	out
13	3
51	41
	8
29	

3. in → Rule → out

in	out
45	25
21	1
70	
	37

4. in → Rule → out

in	out
12	32
28	48
30	
	65

Make up your own.

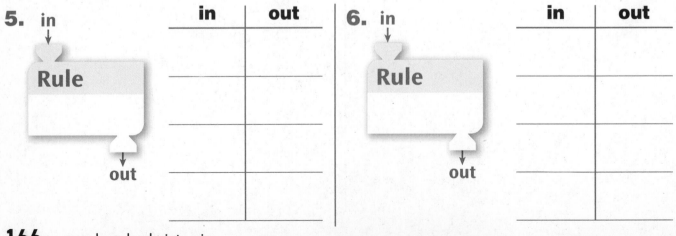

5. in → Rule → out

in	out

6. in → Rule → out

in	out

LESSON 8·6 Math Boxes

1. A ring costs 20¢.

I pay a ⓠ.

How much change will
I get?

_____ ¢

2. How many equal parts?

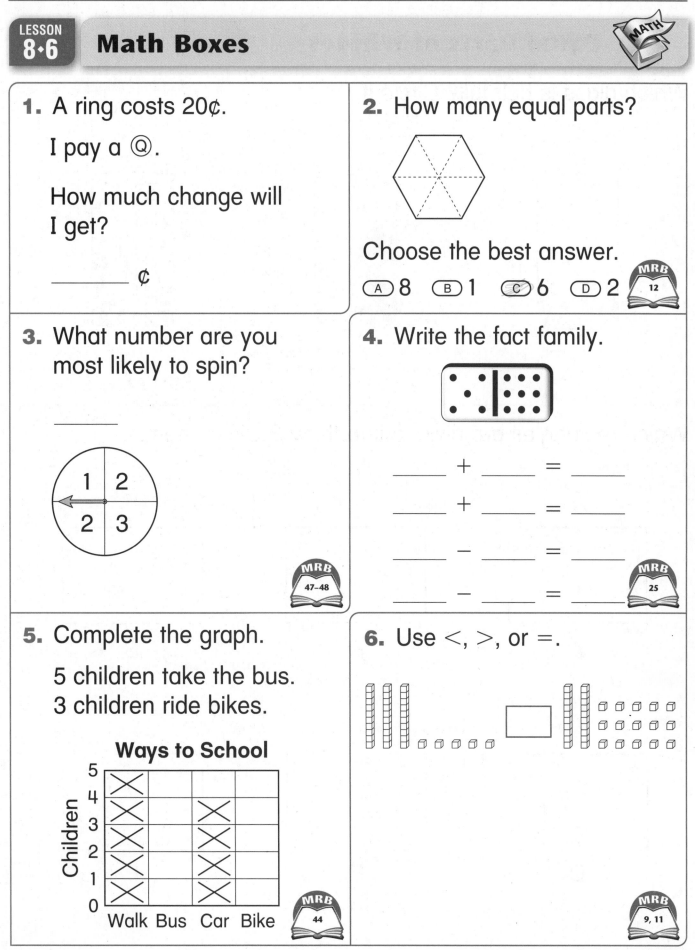

Choose the best answer.

(A) 8   (B) 1   (C) 6   (D) 2   MRB 12

3. What number are you
most likely to spin?

_____

MRB 47–48

4. Write the fact family.

_____ + _____ = _____

_____ + _____ = _____

_____ − _____ = _____

_____ − _____ = _____

MRB 25

5. Complete the graph.

5 children take the bus.
3 children ride bikes.

Ways to School

MRB 44

6. Use <, >, or =.

MRB 9, 11

LESSON 8·7 Equal Parts of Wholes

Which glass is half full? Circle it.

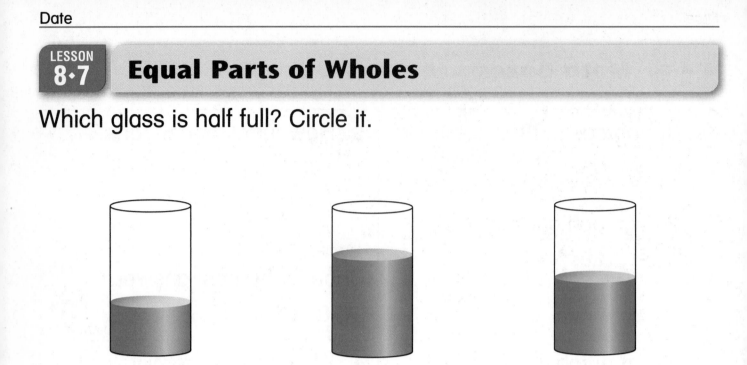

Which rectangles are divided into thirds? Circle them.

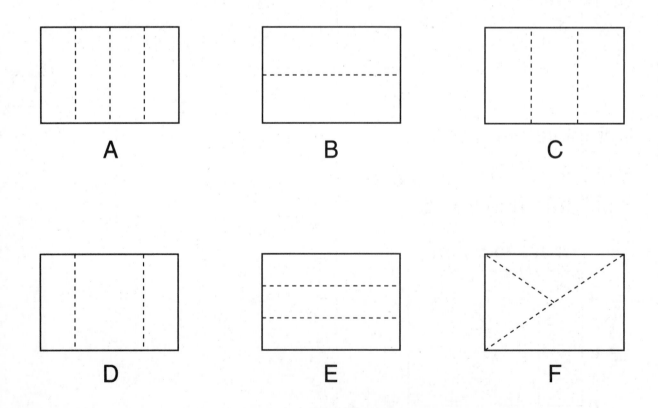

A   B   C

D   E   F

LESSON 8·7 Fractions

1. How many equal parts are there? _____

   Write a fraction in each part of the circle.

   Color $\frac{1}{3}$ of the circle.

2. How many equal parts are there? _____

   Write a fraction in each part of the square.

   Color $\frac{1}{4}$ of the square.

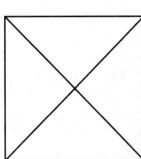

3. How many equal parts are there? _____

   Write a fraction in each part of the hexagon.

   Color $\frac{1}{6}$ of the hexagon.

4. How many equal parts are there? _____

   Write a fraction in each part of the rectangle.

   Color $\frac{1}{8}$ of the rectangle.

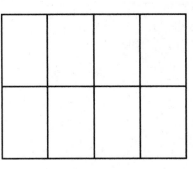

LESSON 8·7   Math Boxes

1. Circle the hundreds place.

289   300   112

733   999   205

MRB 10

2. Carlos bought 3 pencils.
Each pencil costs 10¢.
How much did Carlos pay?

_____ ¢ or $_____._____

3. Draw the other half.

MRB 60

4. Use a straightedge.

Draw line segments to make a polygon.

MRB 52–53

5. Label the box.
Add 3 new names.

Ⓝ Ⓟ Ⓟ Ⓟ

18 − 10

MRB 16

6. Subtract.

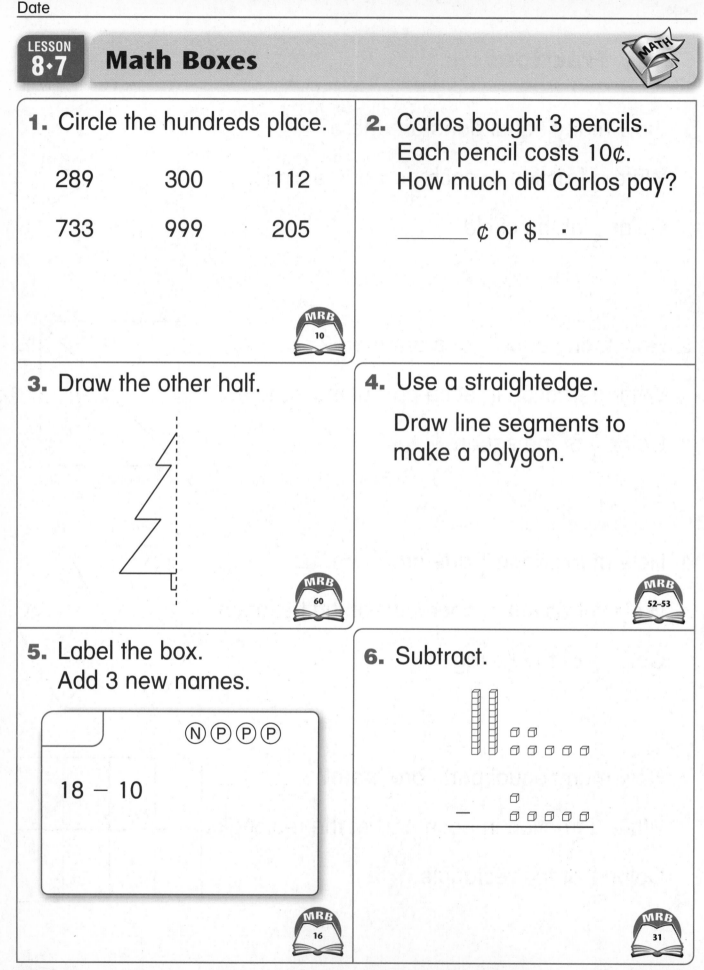

MRB 31

Date _____

LESSON 8·8   **Sharing Pennies**

Use your pennies to help you solve the problems.

Circle each person's share.

1. Halves: 2 people share 8 pennies equally.

   How many pennies does each person get? _____ pennies

2. Thirds: 3 people share 9 pennies equally.

   How many pennies does each person get? _____ pennies

   How many pennies do 2 of the 3 people get in all?

   _____ pennies

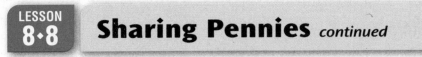

LESSON 8·8   **Sharing Pennies** *continued*

3. Fifths: 5 people share 15 pennies equally.

How many pennies does each person get? _____ pennies

How many pennies do 3 of the 5 people get in all?

_____ pennies

4. Fourths: 4 people share 20 pennies.

How many pennies does each person get? _____ pennies

How many pennies do 2 of the 4 people get in all?

_____ pennies

Date

1. A seashell costs $0.48.
   I pay 2 ⓠ.

   How much change will
   I get?

   _____ ¢

2. Label each equal part.

   MRB
   12–13

3. Complete the graph.

   5 children live 6 blocks away.
   4 children live 7 blocks away.

   Blocks from School

   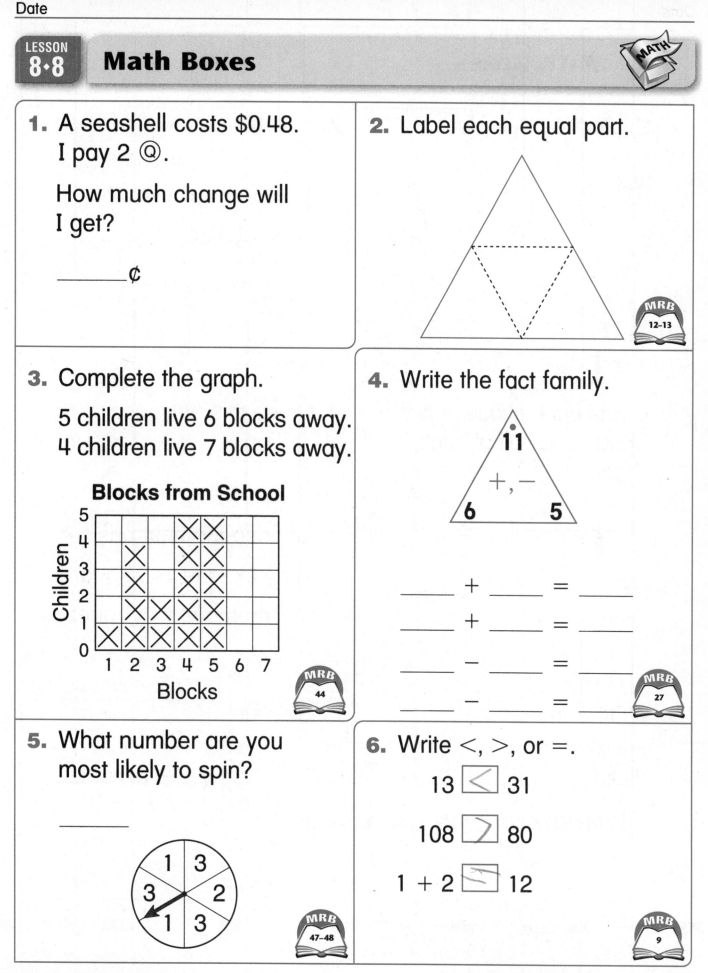

   MRB
   44

4. Write the fact family.

   11
   +, −
   6   5

   ____ + ____ = ____

   ____ + ____ = ____

   ____ − ____ = ____

   ____ − ____ = ____

   MRB
   27

5. What number are you
   most likely to spin?

   _____

   MRB
   47–48

6. Write <, >, or =.

   13 ☐< 31

   108 ☐> 80

   1 + 2 ☐ 12

   MRB
   9

LESSON 8·9 Math Boxes

1. Circle the ones place.

| 364 | 58 | 100 |
| 4 | 16 | 222 |

MRB 10

2. You buy 2 packs of seeds. Each pack costs 60¢.

How much do you pay?

_____ ¢ or $_____ · _____

3. Divide each shape in half. Shade one half of each shape.

MRB 12

4. Name this polygon.

Choose the best answer.

ⓐ hexagon ⓑ square

ⓒ rhombus ⓓ trapezoid

MRB 54–55

5. Write 3 more names.

100

80 + 20

MRB 16

6. Subtract.

MRB 31

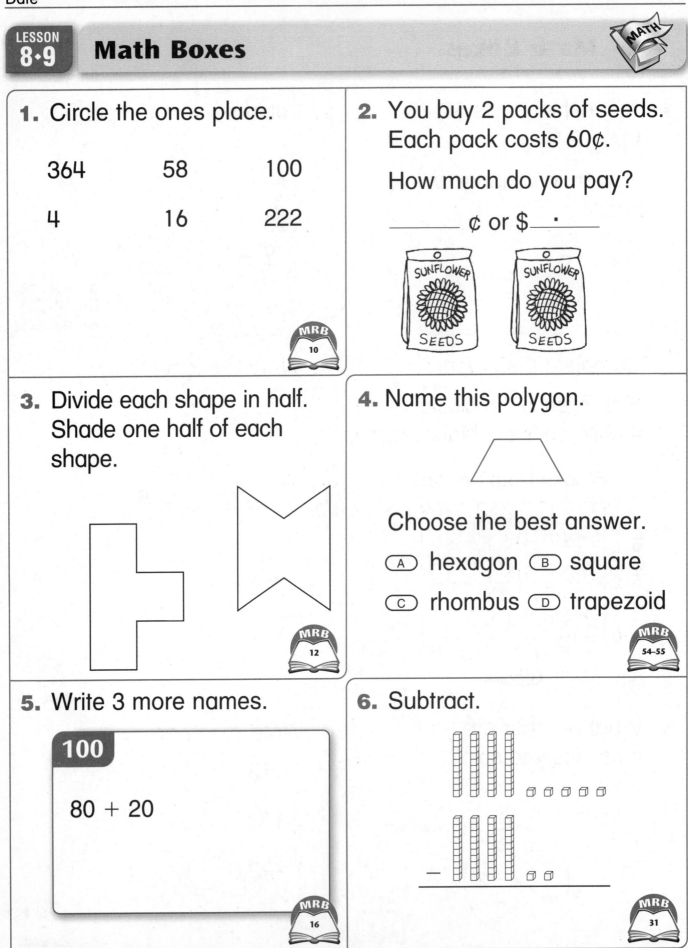

174 one hundred seventy-four

Date _____

1. Use your number grid.
Start at 26.
Count up 14.

$$\begin{array}{r} 26 \\ + \ 14 \\ \hline \end{array}$$

MRB
29

2. Shade $\frac{1}{4}$ of the circle.

MRB
13

3. Fill in the missing numbers.

in ↓

Rule
Add 5

out

in	out
8	
12	
29	
41	
100	

MRB
100–102

4. Write the fact family.

10
+, −
7   3

_____ + _____ = _____

_____ + _____ = _____

_____ − _____ = _____

_____ − _____ = _____

MRB
27

5. Tell the time.

_____ : _____

MRB
81

6. Freddy has Ⓠ Ⓓ Ⓓ Ⓝ.
Jewel has Ⓓ Ⓓ Ⓝ Ⓓ Ⓓ Ⓟ.

Who has more money?

_____

How much more money?

_____ ¢

LESSON 9·2   *Number-Grid Game*

Materials
- ☐ a number grid
- ☐ a die
- ☐ a game marker for each player

Players   2 or more

Skill   Counting on the number grid

Object of the Game   To land on 110 with an exact roll

1. Players put their markers at 0 on the number grid.

2. Take turns. When it is your turn:
   - ◆ Roll the die.
   - ◆ Use the table to see how many spaces to move your marker.
   - ◆ Move your marker that many spaces.

3. Continue playing. The winner is the first player to get to 110 with an exact roll.

Roll	Spaces
•	1 or 10
••	2 or 20
•••	3
••••	4
•••••	5
••••••	6

LESSON 9·2 Math Boxes

1. Use your number grid.
Start at 90.
Count back 25.

$$\begin{array}{r} 90 \\ -\ 25 \\ \hline \end{array}$$

MRB
32

2. Find the sums.

2 + 5 = _____

12 + 5 = _____

42 + 5 = _____

102 + 5 = _____

3. Shade $\frac{1}{2}$ of the pennies.

Ⓟ Ⓟ Ⓟ Ⓟ Ⓟ

Ⓟ Ⓟ Ⓟ Ⓟ Ⓟ

MRB
14

4. Asha bought a key chain for 43¢.

She paid Ⓠ Ⓠ.

How much change did she get? _____ ¢

Show this amount with Ⓓ, Ⓝ, and Ⓟ.

5. Fill in the missing numbers.

MRB
98, 99

Rule
−1

| 653 | 652 | | | | |

6. Circle the four polygons.

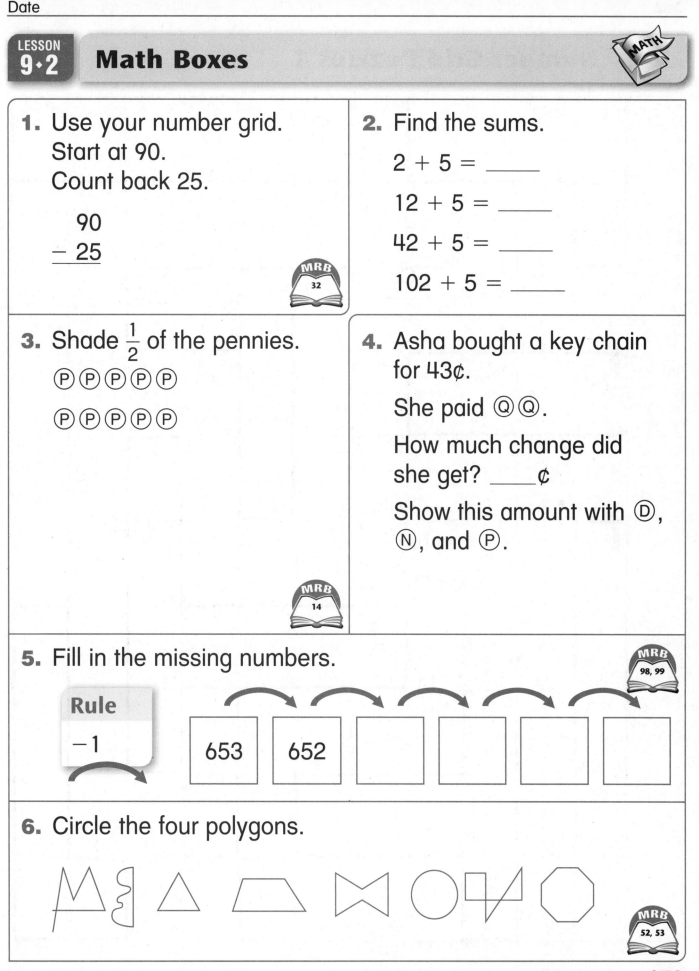

MRB
52, 53

LESSON 9·3 Number-Grid Puzzles 1

	0						60					
	9											
	7											
							99					
			44									
	2											
		21				61				91		

LESSON 9·3

Math Boxes

1. Use your number grid.
Start at 36.
Count up 22.

36 + 22 = _____

MRB 29

2. Divide the rhombus in half.
Shade $\frac{1}{2}$.

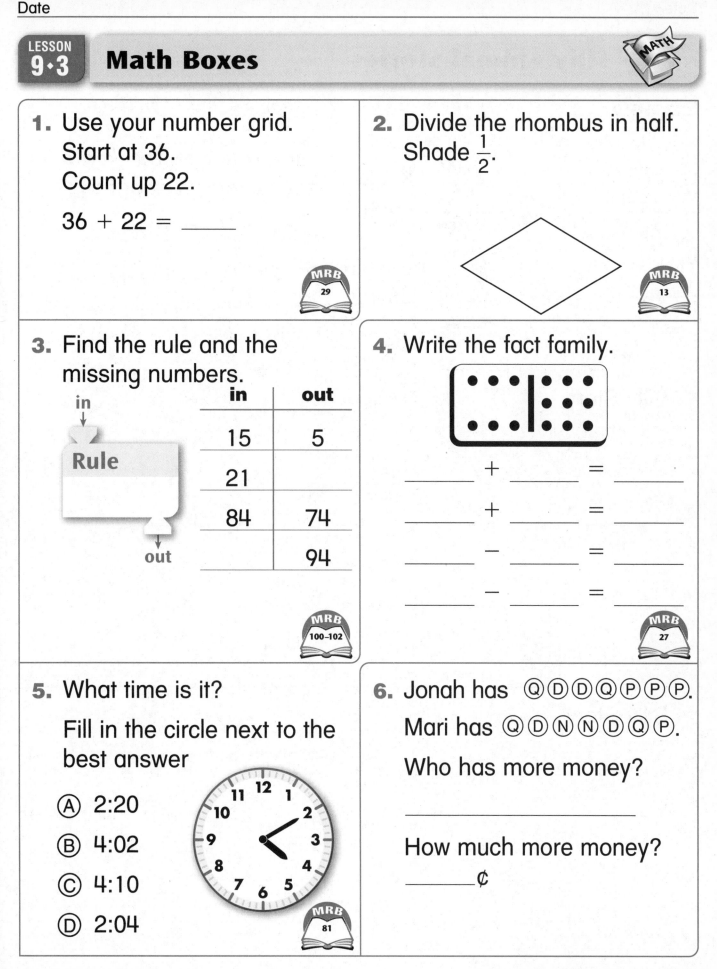

MRB 13

3. Find the rule and the missing numbers.

in ↓

Rule

out ↓

in	out
15	5
21	
84	74
	94

MRB 100–102

4. Write the fact family.

_____ + _____ = _____

_____ + _____ = _____

_____ − _____ = _____

_____ − _____ = _____

MRB 27

5. What time is it?

Fill in the circle next to the best answer

Ⓐ 2:20

Ⓑ 4:02

Ⓒ 4:10

Ⓓ 2:04

MRB 81

6. Jonah has Ⓠ Ⓓ Ⓓ Ⓠ Ⓟ Ⓟ Ⓟ.

Mari has Ⓠ Ⓓ Ⓝ Ⓝ Ⓓ Ⓠ Ⓟ.

Who has more money?

_____

How much more money?

_____¢

LESSON 9·4   Silly Animal Stories

Example:

Unit
inches

koala
24 in.

penguin
36 in.

How tall are the koala and penguin together?

$24 + 36 = 60$

60 inches

---

1. Silly Story

Unit

---

2. Silly Story

Unit

LESSON 9·4 **Math Boxes**

1. Use your number grid.
Start at 48. Count back 15.
48 − 15 = ?
Fill in the circle next to the
best answer.

Ⓐ 43   Ⓑ 33

Ⓒ 63   Ⓓ 36

MRB 32

2. Solve.

16 − 9 = _____

26 − 9 = _____

56 − 9 = _____

106 − 9 = _____

3. Draw 12 dimes.
Use Ⓓs.

Shade $\frac{1}{4}$ of the dimes.

MRB 14

4. A toy dinosaur costs 89¢.

I paid $1.00.

How much change do
I get?

_____¢

Show this amount with Ⓓ,
Ⓝ, and Ⓟ.

5. Find the rule. Fill in the missing numbers.

MRB 98, 99

Rule

| 165 | 265 | 365 | | |

6. Circle the 3 polygons.

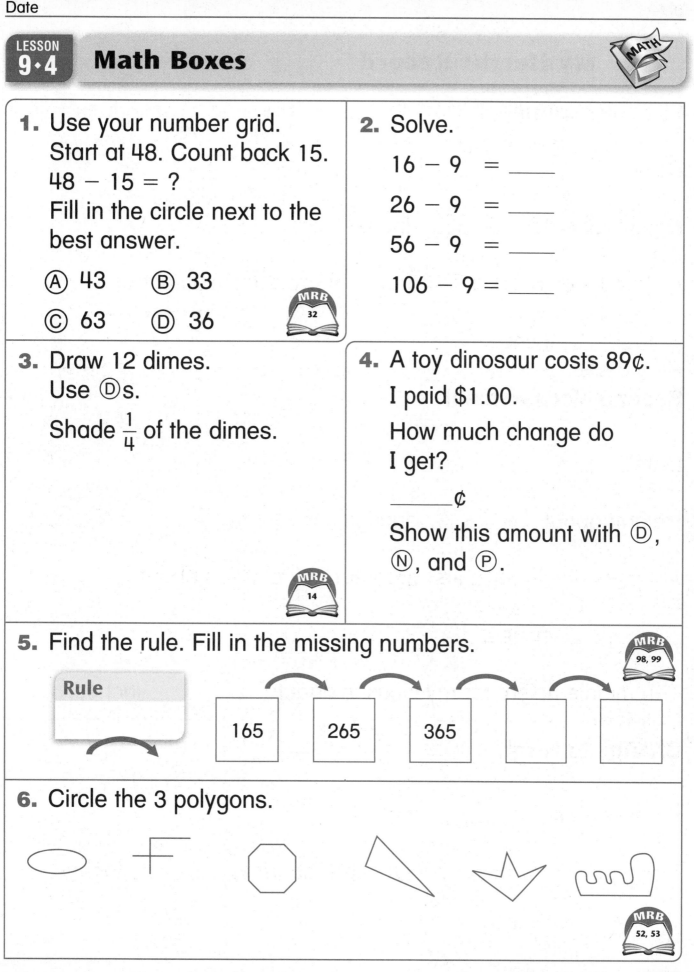

MRB 52, 53

LESSON 9·5   My Height Record

First Measurement

Date _____

Height: about _____ inches

A typical height for a first grader in my class was about

_____ inches.

Second Measurement

Date _____

Height: about _____ inches

A typical height for a first grader in my class is about

_____ inches.

The middle height for my class is about _____ inches.

Change to Height

I grew about _____ inches.

The typical growth in my class was about _____ inches.

LESSON 9·5 Math Boxes

1. Find the sums.

2 + 6 = _____

20 + 60 = _____

200 + 600 = _____

2. Complete the number-grid puzzle.

42

MRB
7

3. Label each part.

Shade $\frac{1}{3}$ of the rectangle.

MRB
13

4. Circle the 4 letters that are symmetrical.

A   P   H   E

L   V   F   Q

MRB
60

5. Myla buys 2 items at the store.

45¢   TOOTHPASTE   39¢

How much money does she spend?

_____

Show this amount with
Ⓠ, Ⓓ, Ⓝ, and Ⓟ.

6. This is a picture of a 3-D shape.
Circle the name of the shape.

pyramid   cone

MRB
56, 57

LESSON 9·6
Pattern-Block Fractions

Use pattern blocks to divide each shape into equal parts.
Draw the parts using your Pattern-Block Template.
Shade parts of the shapes.

1. Divide the rhombus into halves. Shade $\frac{1}{2}$ of the rhombus.

2. Divide the trapezoid into thirds. Shade $\frac{2}{3}$ of the trapezoid.

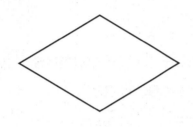

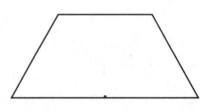

3. Divide the hexagon into halves. Shade $\frac{2}{2}$ of the hexagon.

4. Divide the hexagon into thirds. Shade $\frac{2}{3}$ of the hexagon.

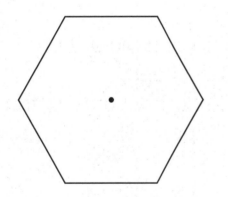

5. Divide the hexagon into sixths. Shade $\frac{4}{6}$ of the hexagon.

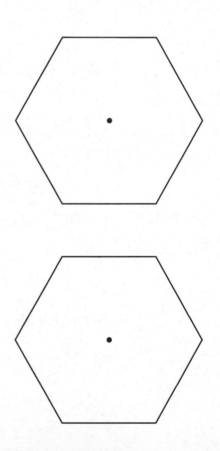

Math Boxes

1. Solve.

_____ = 9 + 9

_____ = 90 + 90

| 7 | 70 | 700 |
| − 5 | − 50 | − 500 |

2. Divide the rectangle into fourths. Shade $\frac{3}{4}$ of the rectangle.

MRB
12, 13

3. Draw and solve.

Griffin had 14 guppies.

He gave $\frac{1}{2}$ away.

How many guppies are left?

_____ guppies

MRB
14

4. Write the numbers.

_____

_____

MRB
10, 11

5. **Weekly Allowance**

Dollars

Erin José Mary Jeff Eliza

MRB
44

Smallest allowance: $_____

Largest allowance: $_____

6. Record the temperature.

°F

_____ °F

Odd or even?

_____

MRB
87, 97

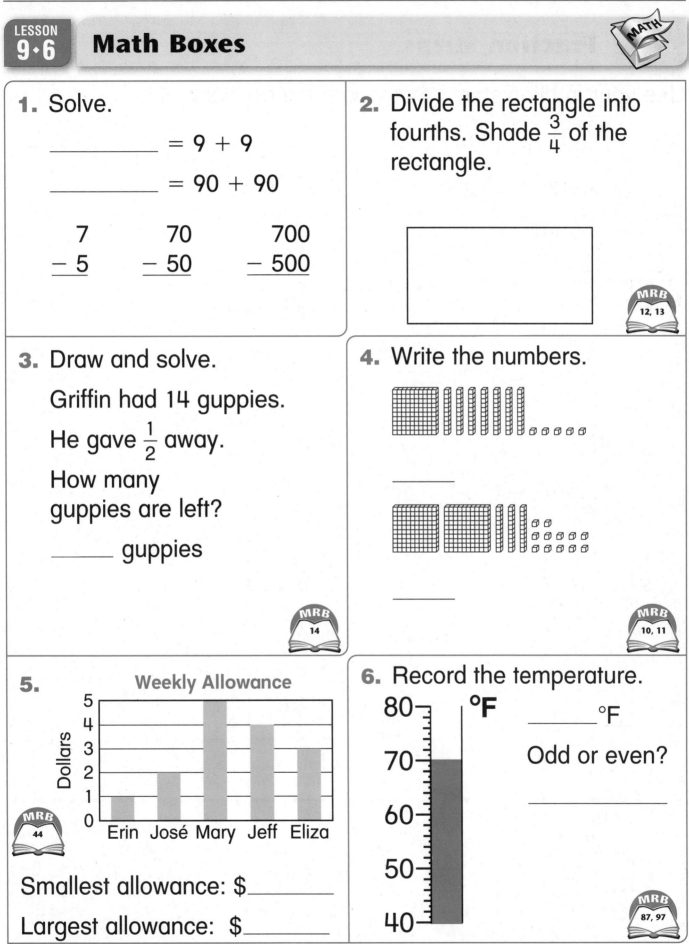

LESSON 9·7 Fraction Strips

Use your fraction strips to compare the fractions.

< is less than
> is greater than
= is equal to

Unit
1-strip

1. $\dfrac{1}{2}$ ☐ $\dfrac{1}{4}$

2. $\dfrac{1}{8}$ ☐ $\dfrac{1}{4}$

3. $\dfrac{1}{2}$ ☐ $\dfrac{1}{8}$

4. $\dfrac{1}{2}$ ☐ $\dfrac{1}{3}$

5. $\dfrac{1}{6}$ ☐ $\dfrac{1}{3}$

6. $\dfrac{1}{4}$ ☐ $\dfrac{1}{3}$

7. $\dfrac{1}{4}$ ☐ $\dfrac{1}{6}$

8. $\dfrac{1}{2}$ ☐ $\dfrac{1}{6}$

Try This

9. $\dfrac{1}{2}$ ☐ $\dfrac{2}{3}$

10. $\dfrac{2}{4}$ ☐ $\dfrac{1}{2}$

LESSON 9·7 **Math Boxes**

1. Solve.

7 − 4 = _____

70 − 40 = _____

700 − 400 = _____

2. Complete the number-grid puzzle.

79

MRB 7

3. Label each part.
Shade $\frac{5}{6}$ of the hexagon.

MRB 13

4. Circle the 3 numbers that are symmetrical.

1 6 3

0 5 4

MRB 60

5. Diego buys 2 items at the store.

DRAWING PAPER

$1.50 CRAYONS 29¢

How much money does he spend?
$_____
Show this amount with $1,
Q, D, N, and P.

6. This is a picture of a 3-dimensional shape. Name the shape.

Fill in the circle next to the best answer.

Ⓐ sphere   Ⓑ cube

Ⓒ cylinder   Ⓓ cone

MRB 56, 57

LESSON 9·8   Many Names for Fractions

1-strip

Use your fraction pieces to help you solve the following problems.

Example:

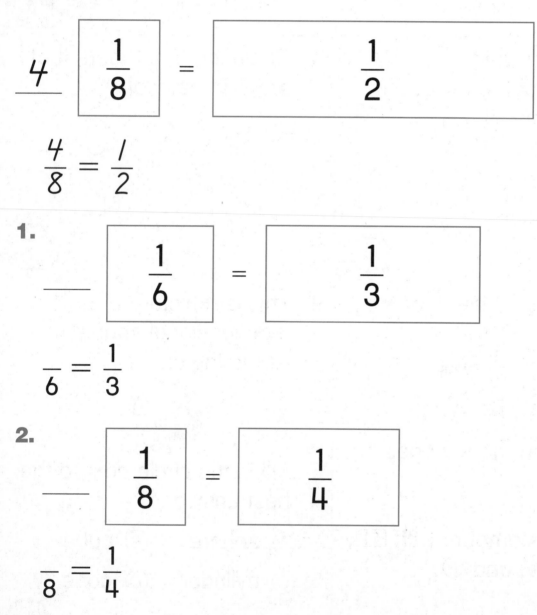

$\dfrac{4}{}$ $\boxed{\dfrac{1}{8}}$ = $\boxed{\dfrac{1}{2}}$

$\dfrac{4}{8} = \dfrac{1}{2}$

1.

$\dfrac{}{}$ $\boxed{\dfrac{1}{6}}$ = $\boxed{\dfrac{1}{3}}$

$\dfrac{}{6} = \dfrac{1}{3}$

2.

$\dfrac{}{}$ $\boxed{\dfrac{1}{8}}$ = $\boxed{\dfrac{1}{4}}$

$\dfrac{}{8} = \dfrac{1}{4}$

LESSON
9·8

Many Names for Fractions *continued*

3.

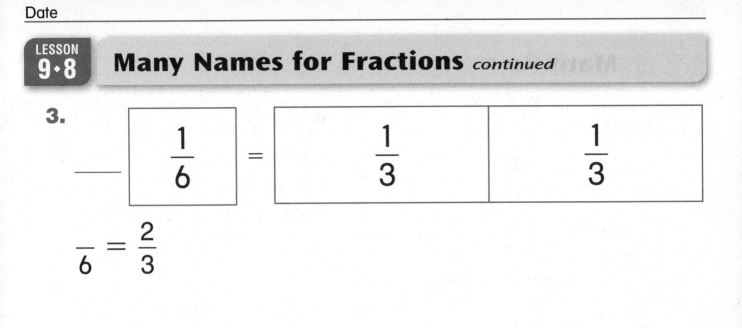

$$\frac{\ }{6} = \frac{2}{3}$$

4.

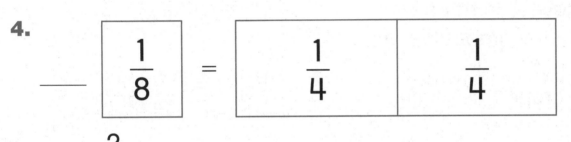

$$\frac{\ }{8} = \frac{2}{4}$$

5.

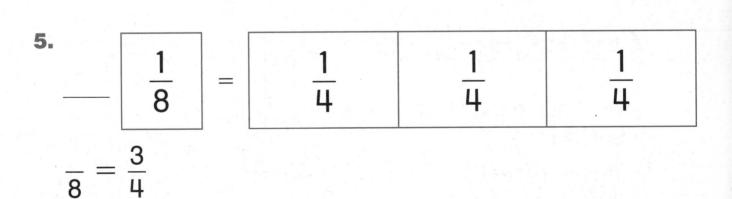

$$\frac{\ }{8} = \frac{3}{4}$$

LESSON 9·8

Math Boxes

1. Solve.

_____ = 9 − 5

_____ = 90 − 50

$$\begin{array}{ccc} 6 & 60 & 600 \\ +\ 4 & +\ 40 & +\ 400 \end{array}$$

2. What fraction is shaded?
Fill in the circle next to the best answer.

Ⓐ $\frac{1}{3}$ Ⓑ $\frac{3}{8}$

Ⓒ $\frac{8}{3}$ Ⓓ $\frac{3}{1}$

MRB 13

3. Draw and solve.
Emma had 15 grapes.
She gave $\frac{1}{3}$ to her sister.
How many grapes did her sister get?

_____ grapes

MRB 14

4. Write the numbers.

_____

_____

MRB 10, 11

5. **Books Read in a Week**

Number of Books

Jerry Sam Eli Ben Julia

Fewest number
of books read: _____
Greatest number
of books read: _____
Range: _____

MRB 44, 45

6. Record the temperature.

_____ °F

Odd or even?

_____

°F
80
70
60
50
40

MRB 87, 97

Date _____

1. Lowest count:

_____

Highest count:

_____

Range:

_____

Calculator Counts in 15 Seconds

Number of Children

6 | | | | | ✕
5 | | | | | ✕
4 | ✕ | | | | ✕
3 | ✕ | | ✕ | | ✕
2 | ✕ | | | ✕ | ✕
1 | ✕ | | ✕ | ✕ | ✕
0 | ✕ | | ✕ | ✕ | ✕

15 16 17 18 19 20

Counted to

MRB 44, 45

2. Sam buys 2 items at the store.

15¢ $3.25

How much money does he spend?

_____

Show this amount with $1, Ⓠ, Ⓓ, Ⓝ, and Ⓟ.

MRB 88–90

3. You have $1.00. You buy a pretzel that costs 75¢.

How much change do you get?

_____¢

Show this amount with Ⓠ, Ⓓ, Ⓝ, and Ⓟ.

MRB 52–53

4. Pedro has Ⓠ Ⓓ Ⓝ Ⓟ Ⓝ Ⓝ Ⓓ. Claudia has Ⓓ Ⓓ Ⓝ Ⓠ Ⓝ Ⓠ Ⓟ.

Who has more money?

_____

How much more money?

_____¢

5. Use a ruler. Draw a polygon with 4 sides.

6. Draw the hands.

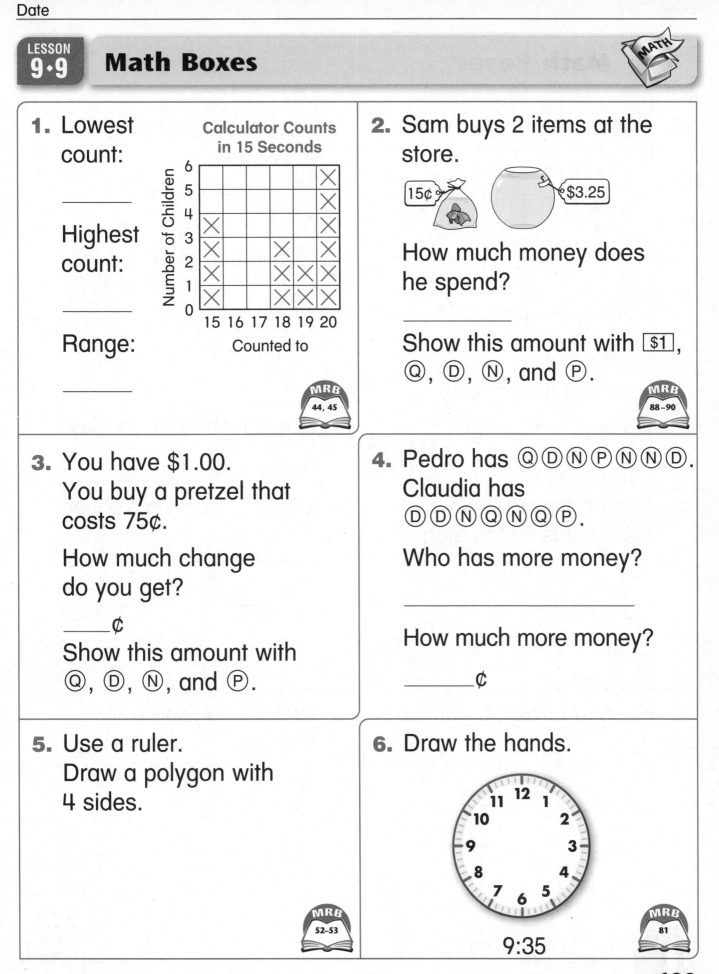

9:35

MRB 81

one hundred ninety-three **193**

LESSON 10·1 Math Boxes

1. Record the temperatures.

Temperatures
Recorded This Month

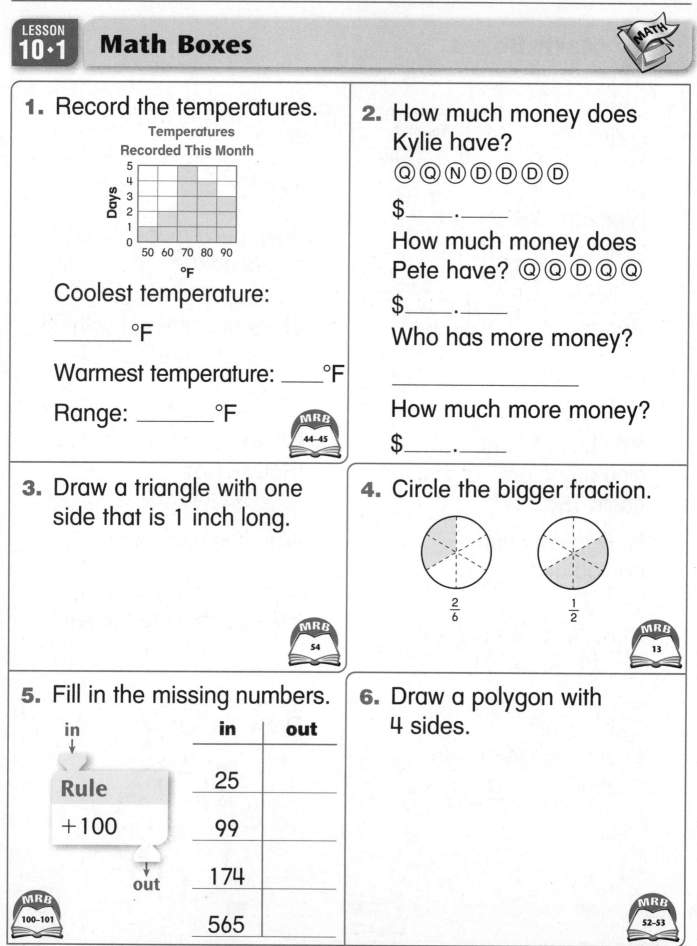

Coolest temperature:

_____°F

Warmest temperature: ____°F

Range: _____°F

MRB
44–45

2. How much money does Kylie have?

Ⓠ Ⓠ Ⓝ Ⓓ Ⓓ Ⓓ Ⓓ

$____.____

How much money does Pete have? Ⓠ Ⓠ Ⓓ Ⓠ Ⓠ

$____.____

Who has more money?

_____

How much more money?

$____.____

3. Draw a triangle with one side that is 1 inch long.

MRB
54

4. Circle the bigger fraction.

$\frac{2}{6}$ $\frac{1}{2}$

MRB
13

5. Fill in the missing numbers.

in ↓

Rule
+100

out ↓

in	out
25	
99	
174	
	565

MRB
100–101

6. Draw a polygon with 4 sides.

MRB
52–53

Date _____

1. Ask a partner to show times on a tool-kit clock.
   Draw the hands on the clock. Write the times to match.

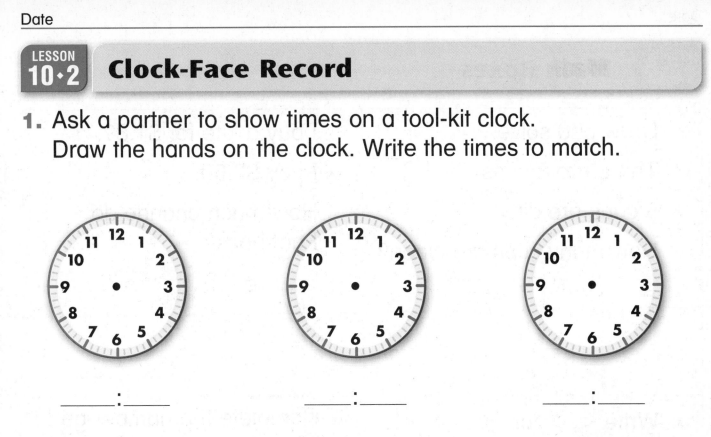

_____:_____   _____:_____   _____:_____

2. Write a time for each clock face. Draw the hands to match.

_____:_____   _____:_____   _____:_____

3. Set your tool-kit clock to 3:00.
   How many minutes until 3:25? _____ minutes

   Set your tool-kit clock to 1:30.
   How many minutes until 1:55? _____ minutes

   Set your tool-kit clock to 10:45.
   How many minutes until 11:20? _____ minutes

LESSON 10·2 Math Boxes

1. Draw and solve.

There are 8 cups.

5 cups are dirty.

How many cups are clean?

_____ cups

2. I buy a kite for $1.89.

I pay $2.00.

How much change do I get back?

_____ ¢

3. Write <, >, or =.

305 ☐ 385

113 ☐ 100 + 13

129 ☐

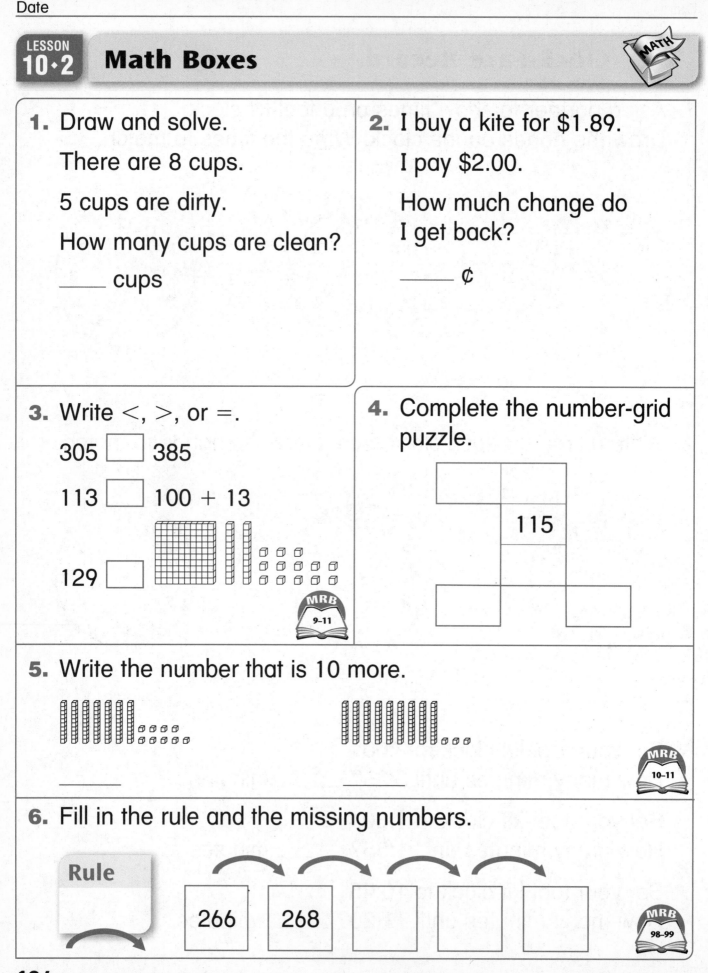

MRB
9–11

4. Complete the number-grid puzzle.

115

5. Write the number that is 10 more.

_____   _____

MRB
10–11

6. Fill in the rule and the missing numbers.

Rule

266 268 ☐ ☐ ☐

MRB
98–99

LESSON 10·3 Vending Machine Poster

LESSON 10·3 Buying from the Vending Machine

Pretend to buy items from the vending machine. Draw pictures or write the names of items you buy. Show the coins you use to pay for the items. Use Ⓝ, Ⓓ, and Ⓠ. Write the total cost.

1.

2.

3.

4.

Show the cost of these items. Use Ⓝ, Ⓓ, and Ⓠ.
Write the total cost.

5.

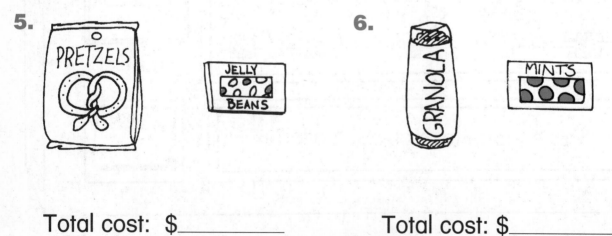

Total cost: $_____

6.

Total cost: $_____

LESSON 10·3 Math Boxes

1. Record the times.

First-Grade Bedtimes

Children
4
3
2
1
0

7:30 8:00 8:30 9:00 9:30

P.M.

Earliest bedtime: _____

Latest bedtime: _____

Range: _____ hours

MRB
44–45

2. Clay has $1 Q D D D.

Rosa has Q Q Q Q Q Q Q.

Who has more money?

_____

How much more money?

_____ ¢

3. Measure the line segment.

It is about _____ inches long.

_____

Fill in the circle next to the best answer.

○ **A.** 9   ○ **B.** 3

○ **C.** $3\frac{1}{2}$ ○ **D.** $8\frac{1}{2}$

MRB
65

4. Circle the bigger fraction.

$\frac{3}{4}$ $\frac{3}{8}$

MRB
13

5. Fill in the missing numbers.

in
↓

Rule

+10

↓
out

in	out
	18
	86
	103
	264

MRB
100–101

6.

How many sides?

_____ sides

How many corners?

_____ corners

MRB
52–53

LESSON 10·4 Math Boxes

1. Draw and solve.

Amelia has 12 checkers.

6 checkers are black.

How many checkers are not black?

_____ checkers

2. A magnet costs $1.39.

Jamal has $1.25.

How much more money does he need?

_____ ¢

3. Write <, >, or =.

Ⓠ Ⓠ Ⓠ Ⓠ Ⓠ ☐ $1.25

Ⓠ Ⓓ Ⓓ Ⓝ Ⓓ ☐ $0.50

Ⓠ Ⓠ Ⓠ Ⓓ Ⓓ Ⓓ ☐ $1.00

MRB
9, 88, 89

4. Complete the number-grid puzzle.

		334

5. Write the number that is 10 more.

_____   _____

MRB
10–11

6. Fill in the rule and the missing numbers.

Rule					
			148	158	

MRB
98–99

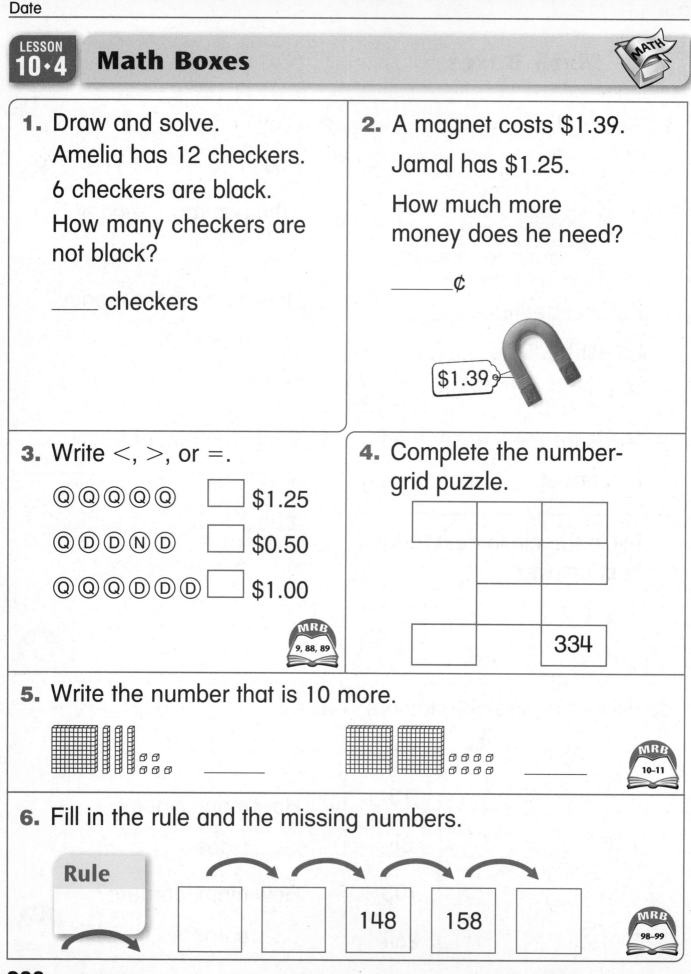

Some Polygons

Triangles

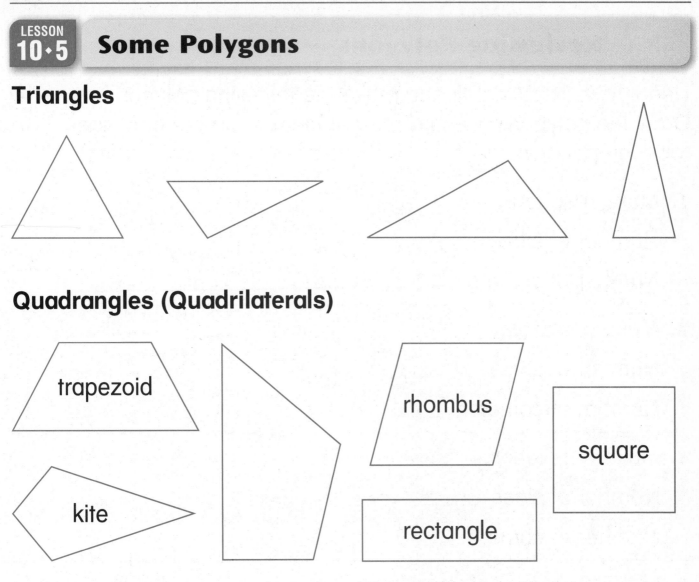

Quadrangles (Quadrilaterals)

trapezoid

kite

rhombus

square

rectangle

Other Polygons

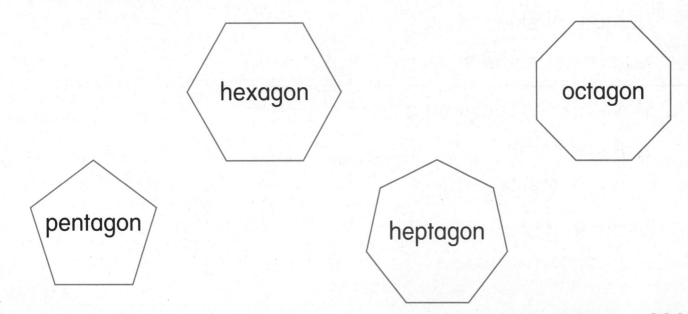

hexagon

octagon

pentagon

heptagon

LESSON 10·5 Reviewing Polygons

Use straws and twist-ties to make the following polygons.
Draw the polygons. Record the number of corners and sides
for each polygon.

1. Make a square.

Number of sides _____

Number of corners _____

2. Make a triangle.

Number of sides _____

Number of corners _____

3. Make a hexagon.

Number of sides _____

Number of corners _____

4. Make a polygon of your choice.

Write its name. _____

Number of sides _____

Number of corners _____

5. Make another polygon of your choice.

Write its name. _____

Number of sides _____

Number of corners _____

Date _____

Word Bank		
sphere	rectangular prism	pyramid
cube	cone	cylinder

Write the name of each 3-dimensional shape.

1.

_____

2.

_____

3.

_____

4.

_____

5.

_____

6.

_____

Five Regular Polyhedrons

The faces that make each shape are identical.

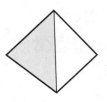

tetrahedron
4 faces

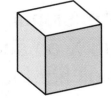

cube
6 faces

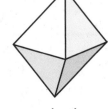

octahedron
8 faces

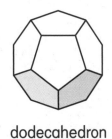

dodecahedron
12 faces

icosahedron
20 faces

LESSON 10·5 **Math Boxes**

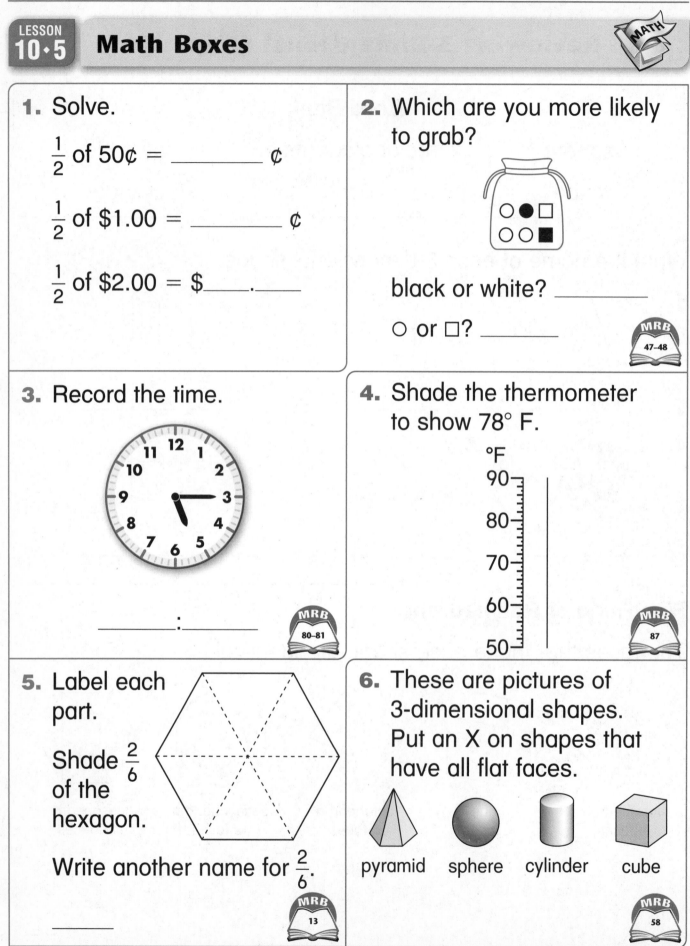

1. Solve.

$\frac{1}{2}$ of 50¢ = _____ ¢

$\frac{1}{2}$ of $1.00 = _____ ¢

$\frac{1}{2}$ of $2.00 = $____.____

2. Which are you more likely to grab?

black or white? _____

○ or □? _____

MRB 47–48

3. Record the time.

_____:_____

MRB 80–81

4. Shade the thermometer to show 78° F.

°F
90
80
70
60
50

MRB 87

5. Label each part.

Shade $\frac{2}{6}$ of the hexagon.

Write another name for $\frac{2}{6}$.

_____

MRB 13

6. These are pictures of 3-dimensional shapes. Put an X on shapes that have all flat faces.

pyramid   sphere   cylinder   cube

MRB 58

U.S. Weather Map

U.S. Weather Map: Spring High/Low Temperatures (°F)

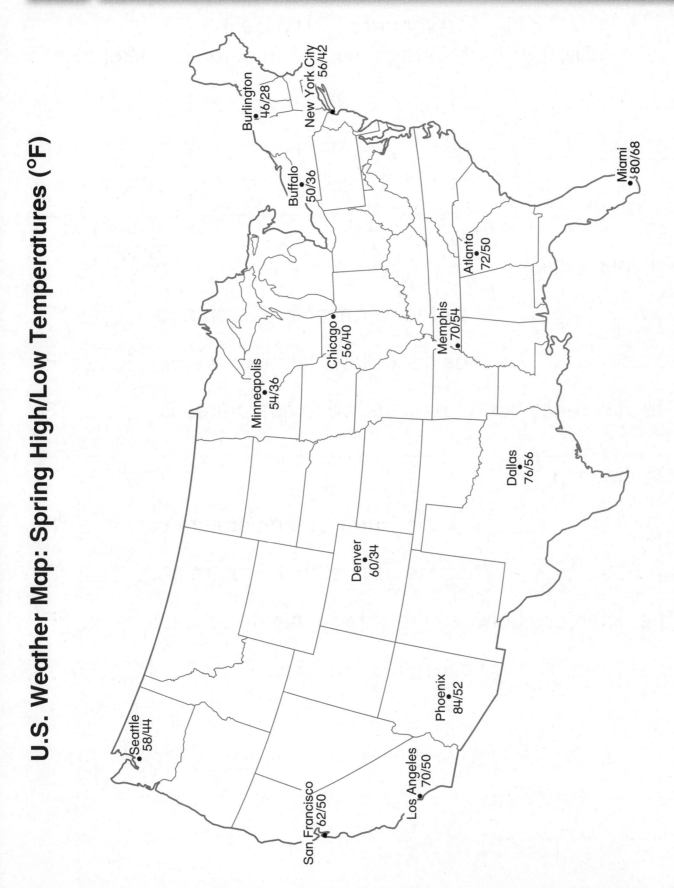

Seattle 58/44

San Francisco 62/50

Los Angeles 70/50

Phoenix 84/52

Denver 60/34

Minneapolis 54/36

Chicago 56/40

Dallas 76/56

Memphis 70/54

Atlanta 72/50

Buffalo 50/36

Burlington 46/28

New York City 56/42

Miami 80/68

LESSON 10·6 **Temperature Chart**

1.

City	Warmest Temperature	Coldest Temperature	Difference
_____	_____ °F	_____ °F	_____ °F
_____	_____ °F	_____ °F	_____ °F
_____	_____ °F	_____ °F	_____ °F

2. Of your 3 cities,

_____ has the warmest temperature at _____ °F.

_____ has the coldest temperature at _____ °F.

The difference between these two temperatures is _____ °F.

3. On the map,

_____ has the warmest temperature at _____ °F.

_____ has the coldest temperature at _____ °F.

The difference between these two temperatures is _____ °F.

Date

1. Draw and solve.

Mateo wants to read 8 books.

He has read 2 books.

How many more books does Mateo have to read?

_____ books

2. Sunglasses cost $3.99.

I pay $5.00.

How much change do I get back?

Fill in the circle next to the best answer.

○ **A.** $2.99   ○ **B.** $8.99

○ **C.** $8.00   ○ **D.** $1.01

3. Write <, >, or =.

10 + 23 ☐ 40

18 + 5 ☐ 5 + 18

32 ☐ 51 − 20

Half of 50 ☐ 25

MRB
9

4. Complete the puzzle.

		200

5. Write the number that is 10 less.

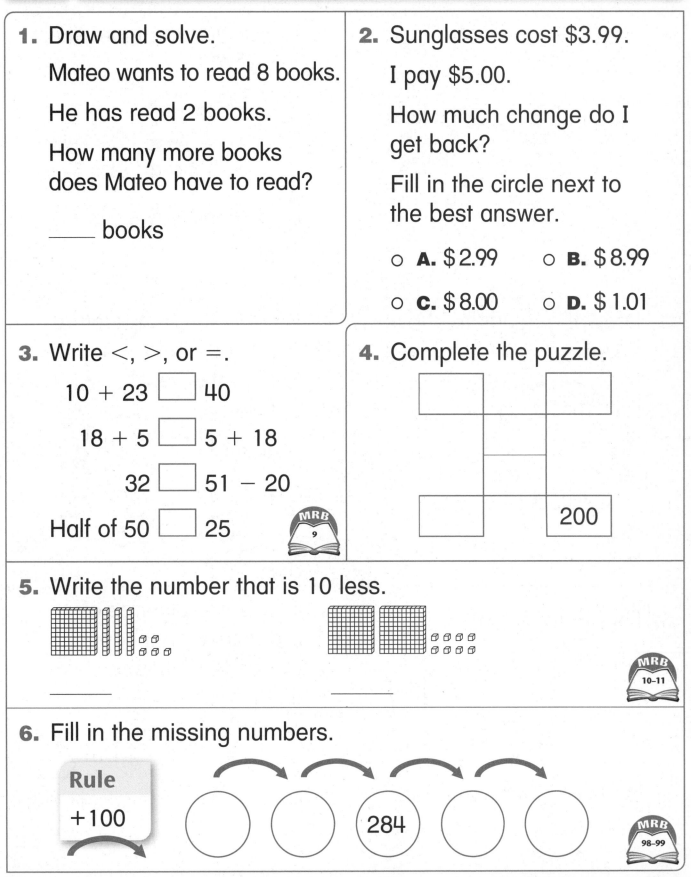

_____   _____

MRB
10–11

6. Fill in the missing numbers.

Rule

+100

( ) ( ) (284) ( ) ( )

MRB
98–99

two hundred seven **207**

LESSON 10·7   Math Boxes

---

1. $2.00 =

_____ pennies

_____ nickels

_____ dimes

_____ quarters

MRB 88–89

2. Which are you more likely to grab?

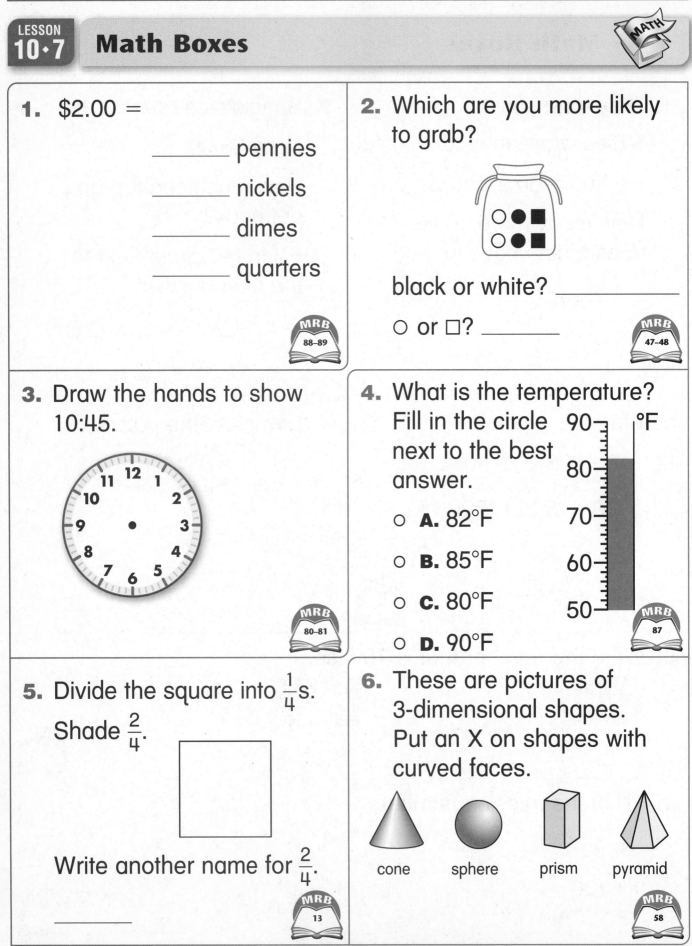

black or white? _____

○ or □? _____

MRB 47–48

---

3. Draw the hands to show 10:45.

MRB 80–81

4. What is the temperature? Fill in the circle next to the best answer.

○ **A.** 82°F

○ **B.** 85°F

○ **C.** 80°F

○ **D.** 90°F

90 ⫟ °F
80
70
60
50

MRB 87

---

5. Divide the square into $\frac{1}{4}$s. Shade $\frac{2}{4}$.

Write another name for $\frac{2}{4}$.

_____

MRB 13

6. These are pictures of 3-dimensional shapes. Put an X on shapes with curved faces.

cone   sphere   prism   pyramid

MRB 58

LESSON 10·8 **Math Boxes**

1. What time is it?

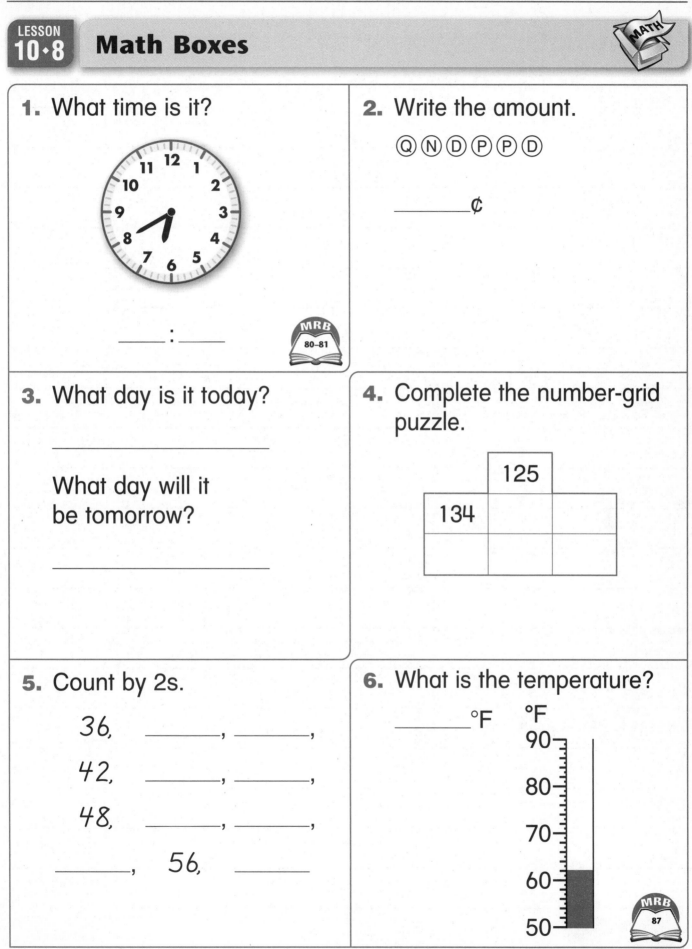

_____ : _____

MRB
80–81

2. Write the amount.

Ⓠ Ⓝ Ⓓ Ⓟ Ⓟ Ⓓ

_____¢

3. What day is it today?

_____

What day will it
be tomorrow?

_____

4. Complete the number-grid puzzle.

	125	
134		

5. Count by 2s.

36, _____, _____,

42, _____, _____,

48, _____, _____,

_____, 56, _____

6. What is the temperature?

_____°F

°F
90
80
70
60
50

MRB
87

Date

Notes

Notes

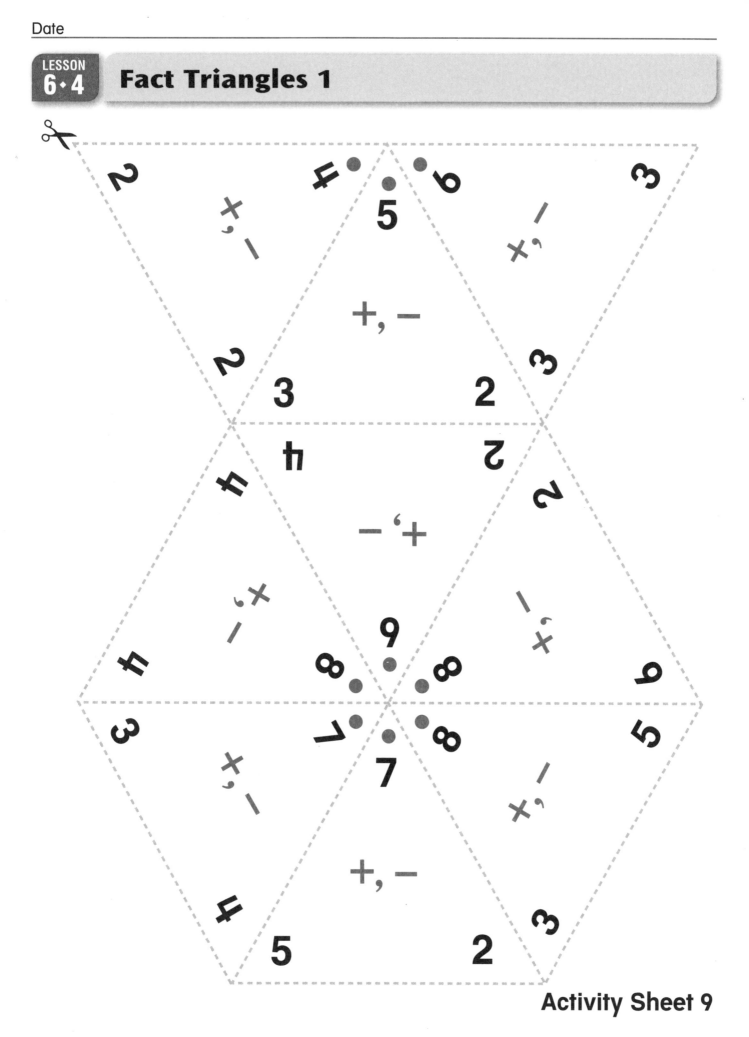

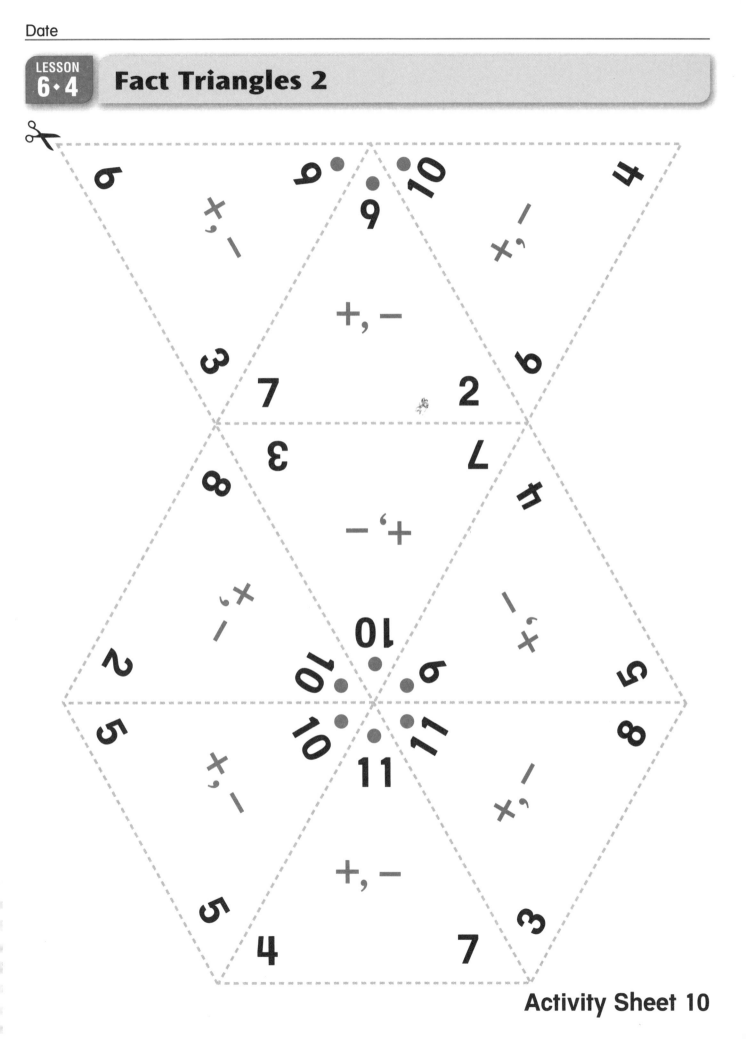

Fact Triangles 3

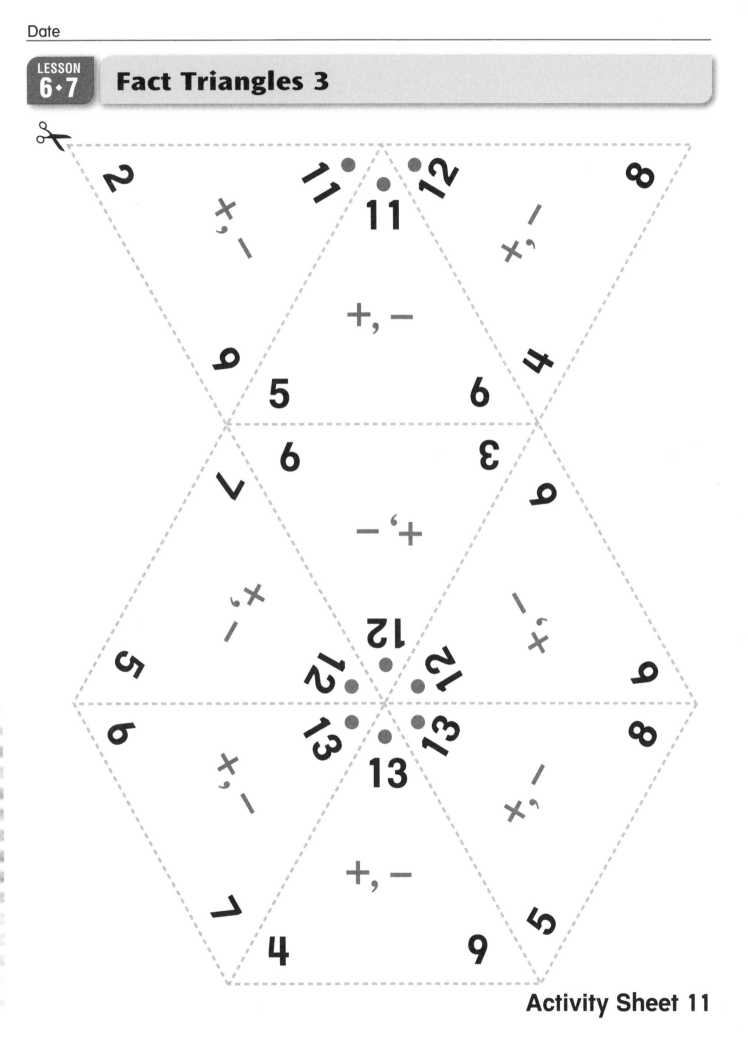

LESSON 6·4 **Fact Triangles 4**

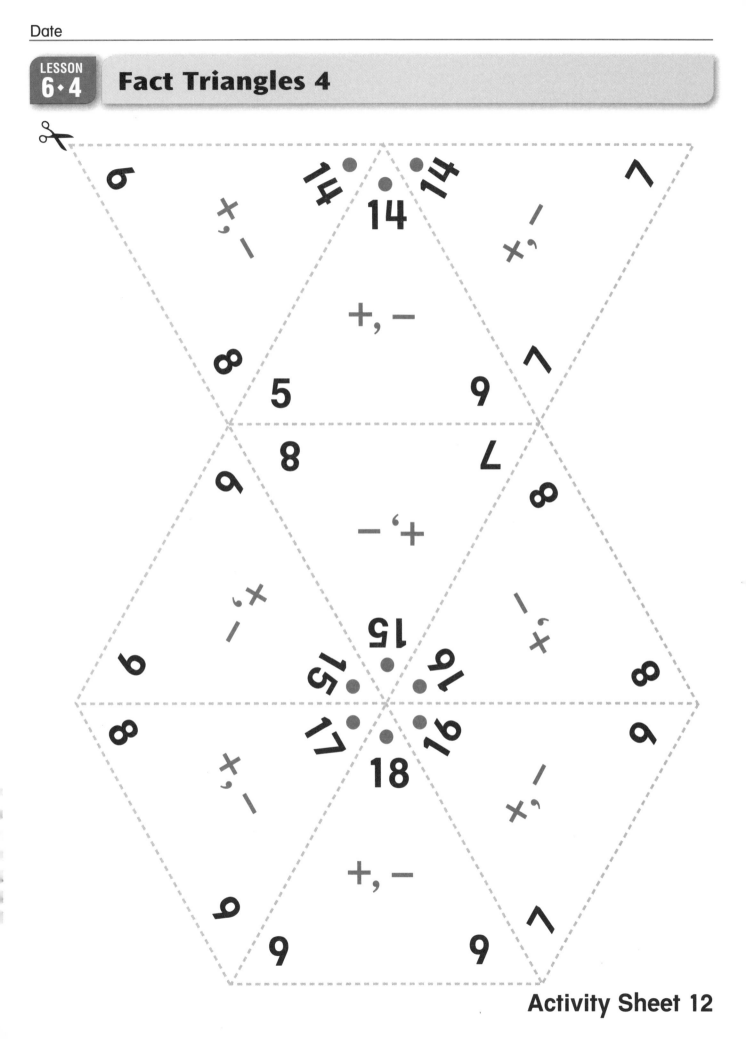

Base-10 Pieces

Base-10 Pieces

Number Grid

0	10	20	30	40	50	60	70	80	90	100	110
-1	9	19	29	39	49	59	69	79	89	99	109
-2	8	18	28	38	48	58	68	78	88	98	108
-3	7	17	27	37	47	57	67	77	87	97	107
-4	6	16	26	36	46	56	66	76	86	96	106
-5	5	15	25	35	45	55	65	75	85	95	105
-6	4	14	24	34	44	54	64	74	84	94	104
-7	3	13	23	33	43	53	63	73	83	93	103
-8	2	12	22	32	42	52	62	72	82	92	102
-9	1	11	21	31	41	51	61	71	81	91	101

Number-Grid Shapes

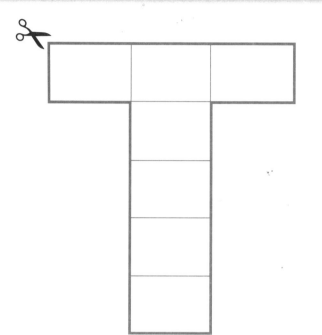

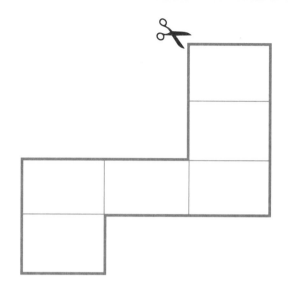

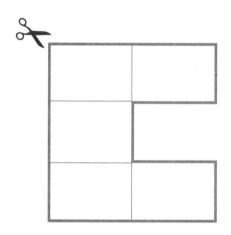

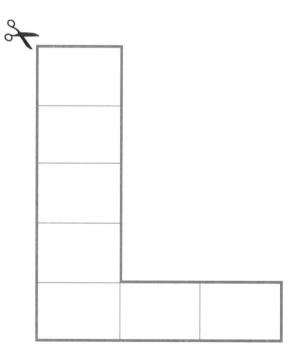

Activity Sheet 16